Advances in Structural Design and Numerical Modelling of Composite Materials

Advances in Structural Design and Numerical Modelling of Composite Materials

Guest Editor

Weilong Yin

MDPI

Basel • Beijing • Wuhan • Barcelona • Belgrade • Novi Sad • Cluj • Manchester

Guest Editor
Weilong Yin
Center for Composite
Materials and Structures
Harbin Institute of
Technology
Harbin
China

Editorial Office
MDPI AG
Grosspeteranlage 5
4052 Basel, Switzerland

This is a reprint of the Special Issue, published open access by the journal *Materials* (ISSN 1996-1944), freely accessible at: https://www.mdpi.com/journal/materials/special_issues/D85DF79S92.

For citation purposes, cite each article independently as indicated on the article page online and as indicated below:

Lastname, A.A.; Lastname, B.B. Article Title. *Journal Name* **Year**, *Volume Number*, Page Range.

ISBN 978-3-7258-6686-1 (Hbk)
ISBN 978-3-7258-6687-8 (PDF)
https://doi.org/10.3390/books978-3-7258-6687-8

Contents

Contents

About the Editor

Weilong Yin

Weilong Yin (1980, 01~), male, associate professor, doctoral supervisor. Graduated from the Aerospace Design program at Beihang University in July 2002, and obtained a Ph.D. in Aerospace Design from Beihang University in July 2007. He joined the Institute of Composite Materials and Structures at Harbin Institute of Technology in July 2007, mainly engaged in research on variant aircraft and flexible adaptive structures. To date, he has published more than 70 papers in well-known domestic and international journals and academic conferences such as *Smart Materials and Structures, Chinese Science, Journal of Aeronautics,* and *Journal of Composite Materials.* Among them, 25 papers have been indexed by SCI, and 60 papers have been indexed by EI; he has submitted 29 national invention patents, and 13 patents have been authorized.

Preface

This Reprint, "Advances in Structural Design and Numerical Modelling of Composite Materials", collects a series of contributions originally published under the same Special Issue title. Its subject is the rapidly evolving field of composite materials engineering, with a specific focus on the integration of advanced structural design principles and sophisticated numerical modelling approaches. The scope of this Reprint spans from fundamental material characterization and failure mechanisms to multiscale simulation techniques and experimental validation, reflecting the interdisciplinary effort required to understand and exploit the complex behaviour of composite systems.

The motivation for compiling this Reprint lies in the growing industrial and academic demand for reliable, efficient, and innovative composite solutions across aerospace, automotive, marine, and civil engineering applications. As composites continue to replace traditional materials in critical load-bearing structures, the ability to accurately predict their performance through simulation-driven design becomes ever more essential. This collection aims to consolidate recent progress and to highlight how numerical tools—from finite element analysis to machine learning-enhanced models—are reshaping the way composites are designed, optimized, and validated.

This Reprint is addressed to researchers, engineers, and postgraduate students active in the fields of composite materials, computational mechanics, and structural design. It offers both a snapshot of current methodologies and a forward-looking perspective on emerging trends in modelling and simulation. By bringing together diverse studies on topics such as damage modelling, interface behaviour, optimization frameworks, and experimental-numerical correlation, the volume seeks to foster further dialogue and innovation in the community.

Weilong Yin
Guest Editor

materials

MDPI

Article

Mechanical Property Analysis of a Boom–Membrane Structure Used for Aerospace Technologies

Shuhong Xu [1,†], Xiaojiao Yu [2,†], Yue Gao [2], Sicong Wang [2,*] and Lining Sun [2]

1 School of Engineering, Applied Technology Collage of Soochow University, Suzhou 215325, China
2 School of Mechanical and Electrical Engineering, Soochow University, Suzhou 215137, China
* Correspondence: scwang8901@suda.edu.cn
† These authors contributed equally to this work.

Abstract: Traditional deployable truss space structures previously had upper limits on their key indicators, such as the deployed area, folded ratio and total weight, and hence, the application of new extendable mechanisms with novel deployment types is desired. Foldable extendable tape spring booms made from FRP (fiber-reinforced polymer) laminate composites and their corresponding boom–membrane structures were invented in recent years to satisfy the needs of the large-scale requirements of spacecraft, especially for antennas, solar sails and solar arrays. This paper aimed to analyze the properties of the deployed states of extendable tape spring booms and their boom–membrane structures. By establishing an analytical model of the boom and the structure, the bending stiffness, critical buckling load of the boom and the fundamental frequency of the membrane structure were acquired. To provide more guidance on the boom–membrane structure design, a geometric and material parametric study was carried out. Meanwhile, an experimental study to investigate the deployed properties of the booms and membrane structures was introduced to afford some practical verification.

Keywords: tape spring; boom–membrane; deployed; aerospace

1. Introduction and Literature Review

Foldable deployable structure technology was previously the main method for solving the envelope limitation issues of large-size spacecraft. Recently, deployable membrane structures were introduced to replace the traditional truss morphing mechanisms in space because of their incomparable advantages such as light weight, high folded ratio and large deployed area. In terms of deployable membrane mechanisms, it was difficult to form smooth and accurate configurations with the inflatable structures after being deployed in orbit, and the ring-shaped truss–membrane structures also had issues with their weight and folded volume when folded. To address this, the invention of boom–membrane structures was a good way to solve these issues, whose deployment and tension-forming process were driven and supported by appropriate elastic extendable mechanisms, for example, tape spring booms [1,2]. Figure 1 presents a diagram of a deployable tape spring boom. An early application of this kind of boom was in the Canadian satellite "Alouette", whose booms were made by Klein in the 1960s [3]. The boom extended outwards like a carpenter's tape measurer, and no hinge was introduced during the whole deployment process, which made this kind of mechanism have a small storage volume after being folded. Under the driving force of the controlling motor connected with the hub, the boom deployed successfully on "Alouette" in space and accomplished the first in-orbit deployment of a tape spring boom. Early deployable booms were mainly made of metal materials, such as steel or CuBe. With the development of the material sciences in the last century, tape spring booms manufactured with composites, especially carbon-/glass-fiber-reinforced polymer (FRP) laminates, were produced and applied [4–7], as the composite laminate materials gave the booms more design flexibility, higher reliability and better deployment stability [8,9].

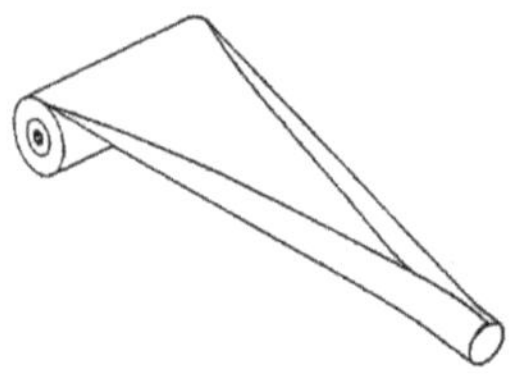

Figure 1. Deployable tape spring boom diagram.

Along with the improvement in this new kind of tape spring boom mechanism, boom–membrane structures that could form planar surface space structures appeared afterward. Boom–membrane structures extremely expanded the application area of tape spring booms, which were further used on large-scale solar arrays, solar sails, observation antennas and space telescope star-shaders. Tape spring booms were also used for a 40 m^2 Synthetic Aperture Radar (SAR) antenna structure developed by DLR and ESA [10–12]. During the SAR's deployment, the beams connected with the boom tips moved outwards under the driving force of the tape spring booms (which were connected by a controlling motor), thus pulling the membrane outwards and stretching and providing tension to the membrane at the end of the deployment. After being fully deployed, the membrane was supported and kept tensioned by the tape spring booms on both sides.

In the 2010s, a prototype of a new boom–membrane solar array was made by the DSS company, called ROSA (Roll-Out Solar Array) because this structure was distinguished from the common types of solar array, like SAR, due to the roll-out deployment type [13,14]. For ROSA, the membrane was coiled on the tip beam when folded and deployed with a roll-out movement of the beam (like opening a scroll), and no controlling motor was needed for the deployment. Because of this, the advantages of the boom–membrane structure were further exploited, and the maximum deployed area was up to 100 m^2 for each unit. DSS also made the structure modularized, and a higher deployed area can be expected. Since the roll-out deployment type could not provide tension to the membrane, a stretching mechanism needed to be arranged at the root of the structures.

Apart from the advantages mentioned above, compared with traditional deployable truss planar structures, the deployed states of boom–membrane structures had a lower stiffness, buckling load and fundamental frequency. The first formal in-orbit application of a boom–membrane structure was on the Hubble telescope's solar array, and the structure buckled and failed soon after launch because of strength and vibration issues [15–17]. The former research was mainly concentrated on the analysis of the deployment process of tape spring booms and membrane structures, and studies on either the deploying or deployed states were relatively independent [18,19]. Since deployable booms have usually been made from laminate composites in recent years, the changes in the laminate parameters would have a relatively large effect on the properties of both the deploying and deployed states. Therefore, following up on the authors' previous research [20,21], this paper establishes an analytical model of tape spring booms and boom–membrane structures in their deployed states, and the parametric effect on their deployed properties is deeply investigated. The multi-configuration optimization of the structures' deploying and deployed states will be further studied in our future work, which will be conducted on the basis of the research in this paper. Section 2 establishes the analytical model of a composite laminate deployable tape spring boom and the corresponding boom–membrane structure, where the boom's bending stiffness, critical buckling load and the fundamental frequency of the membrane structure are acquired. Section 3 analyzes the parametric effect on the boom's buckling load and the membrane system's fundamental frequency to provide more guidance on the boom and the structure design. Further, for the sake of giving some practical verification, the experimental study of the deployed state of the booms and the membrane structures is carried out in Section 4. Section 5 concludes the paper and a discussion is presented in this section.

2. Analytical Model Establishment of a Boom–Membrane Structure

In terms of the boom–membrane structures introduced in Section 1, the tape spring booms were commonly made from composite laminate materials, especially for carbon-fiber- or glass-fiber-reinforced polymers. Therefore, the boom material properties should first be analyzed, and the study of the corresponding boom–membrane structures is investigated afterward in this section.

2.1. Mathematical Model Establishment

For a better understanding of the analysis in this subsection, the diagram of an FRP tape spring boom and the corresponding laminate parameters are given in Figure 2. Generally, fiber-reinforced polymer (FPR) materials consist of one unidirectional (UD) layer in the middle and several fiber layers symmetrically or an-symmetrically distributed on both sides of the UD layer. Note that the model establishment process as follows was carried out under linear hypothesis, and the boom structure was regarded as a thin-walled structure during the whole analysis.

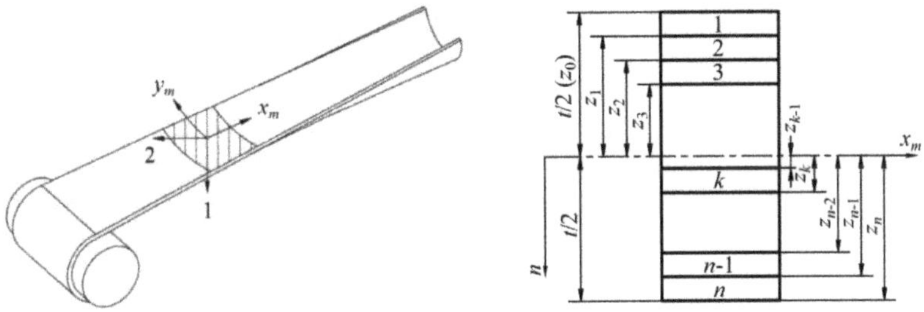

Figure 2. Diagrams of the geometric and laminate parameters of a tape spring boom structure. (**a**) Fiber braided direction (x_m and y_m represent the coordinate system of the boom laminates in which x_m points at the boom's deployment direction). (**b**) Laminate material parameters (the symbols in this figure were the same as those commonly used in the Classical Laminate Theory, which can be found in Ref. [3]).

Firstly, according to the theory of materials mechanics, the elastic parameters of the UD layer can be presented through the equations as follows [8]:

$$\begin{cases} E_{1UD} = E_f V_{UD} + (1-\phi_{UD})E_m(1-V_{UD}) \\ E_{2UD} = \frac{E_f(1-\phi_{UD})E_m}{(1-\phi_{UD})E_m V_{UD} + E_f(1-V_{UD})} \\ \nu_{12UD} = \nu_f V_{UD} + \nu_m(1-V_{UD}) \\ G_{12UD} = \frac{1}{\frac{V_{UD}}{G_f} + \frac{1-V_{UD}}{(1-\phi_{UD})G_m}} \end{cases} \quad (1)$$

where E_{1UD} and E_{2UD} are the elastic moduli along the fiber direction and the normal direction, respectively (see Figure 2a for more details), E_f and E_m are the elastic moduli of the fiber and the matrix, V_{UD} and ϕ_{UD} are the fiber volume friction and the porosity coefficient of the UD layer, ν_{12UD}, ν_f and ν_m are the Poisson's ratios of the UD layer (overall), the fibers and the matrix, and G_{12UD}, G_f and G_m are the shear moduli of the UD layer (overall), the fibers and the matrix, respectively.

Secondly, similarly, the parameters of the fiber layer are acquired only through substituting the subscripts in Equation (1), which can be expressed as:

$$\begin{cases} E_{1f} = E_f V_f + \left(1 - \phi_f\right) E_m \left(1 - V_f\right) \\ E_{2f} = \frac{E_f \left(1-\phi_f\right) E_m}{\left(1-\phi_f\right) E_m V_f + E_f \left(1-V_f\right)} \\ \nu_{12f} = \nu_f V_f + \nu_m \left(1 - V_f\right) \\ G_{12f} = \frac{1}{\frac{V_f}{G_f} + \frac{1-V_f}{\left(1-\phi_f\right) G_m}} \end{cases} \tag{2}$$

According to the Classical Laminate Theory (CLT), the elastic constant of each lamina was obtained through [8] (note that the definitions of the parameters in Equations (1) to (8) were the same as those commonly used in the CLT which would not be re-defined or clarified in this paper. Meanwhile, the corresponding deformation diagrams of FRP composites can be seen in Figure 3 for better understanding):

$$\boldsymbol{Q} = \begin{bmatrix} Q_{11} & Q_{12} & 0 \\ Q_{12} & Q_{22} & 0 \\ 0 & 0 & Q_{66} \end{bmatrix} \tag{3}$$

in which

$$\begin{cases} Q_{11} = \frac{E_1}{1-\nu_{12}\nu_{21}} \\ Q_{22} = \frac{E_2}{1-\nu_{12}\nu_{21}} \\ Q_{12} = \frac{\nu_{12} E_1}{1-\nu_{12}\nu_{21}} \\ Q_{66} = G_{12} \\ \frac{\nu_{12}}{E_2} = \frac{\nu_{21}}{E_1} \end{cases} \tag{4}$$

and the angle conversion formula was

$$\boldsymbol{T} = \begin{bmatrix} \cos^2\delta & \sin^2\delta & 2\sin\delta\cos\delta \\ \sin^2\delta & \cos^2\delta & -2\sin\delta\cos\delta \\ -\sin\delta\cos\delta & \sin\delta\cos\delta & \cos^2\delta - \sin^2\delta \end{bmatrix} \tag{5}$$

where δ is the fiber braided angle in each lamina.

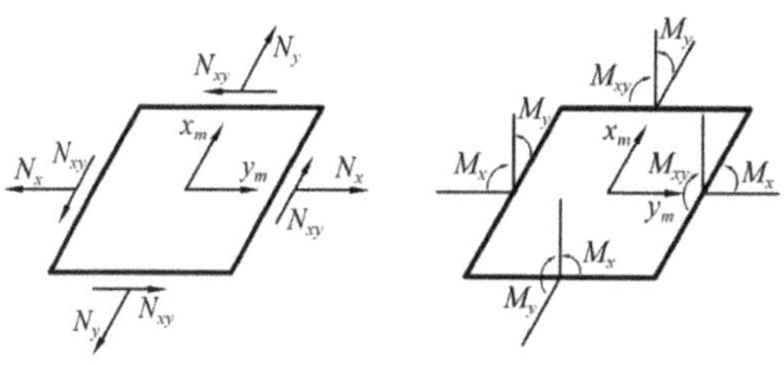

(**a**) Stretching stress (**b**) Bending stress

Figure 3. Stress diagrams of laminate composites in deformation (coordinate $x_m O y_m$ is the same with that listed in Figure 2a).

Therefore, the elastic constant of each lamina after the angle converted can be finally expressed as:

$$\overline{\boldsymbol{Q}} = \begin{bmatrix} \overline{Q}_{11} & \overline{Q}_{12} & \overline{Q}_{16} \\ \overline{Q}_{12} & \overline{Q}_{22} & \overline{Q}_{26} \\ \overline{Q}_{16} & \overline{Q}_{26} & \overline{Q}_{66} \end{bmatrix} = \boldsymbol{T}^{-1}\boldsymbol{Q}\left(\boldsymbol{T}^{-1}\right)^T \tag{6}$$

Moreover, based on the elastic property of each lamina presented in Equation (6), the elastic behaviors of overall laminate composites can be described by the ABD matrix of the CLT, which is commonly shown as:

$$\begin{bmatrix} \boldsymbol{N} \\ \boldsymbol{M} \end{bmatrix} = \begin{bmatrix} \boldsymbol{A} & \boldsymbol{B} \\ \boldsymbol{B} & \boldsymbol{D} \end{bmatrix} \begin{bmatrix} \boldsymbol{\varepsilon} \\ \boldsymbol{\kappa} \end{bmatrix} \tag{7}$$

which might be expressed more specifically as:

$$\begin{bmatrix} N_x \\ N_y \\ N_{xy} \\ M_x \\ M_y \\ M_{xy} \end{bmatrix} = \begin{bmatrix} A_{11} & A_{12} & A_{16} & B_{11} & B_{12} & B_{16} \\ A_{12} & A_{22} & A_{26} & B_{12} & B_{22} & B_{26} \\ A_{16} & A_{26} & A_{66} & B_{16} & B_{26} & B_{66} \\ B_{11} & B_{12} & B_{16} & D_{11} & D_{12} & D_{11} \\ B_{12} & B_{22} & B_{26} & D_{12} & D_{22} & D_{26} \\ B_{16} & B_{26} & B_{66} & D_{16} & D_{26} & D_{66} \end{bmatrix} \begin{bmatrix} \varepsilon_x \\ \varepsilon_y \\ \gamma_{xy} \\ \kappa_x \\ \kappa_y \\ \kappa_{xy} \end{bmatrix} \tag{8}$$

Furthermore, the elements in the ABD matrix in Equation (8) can be acquired by:

$$\begin{cases} A_{ij} = \sum\limits_{k=1}^{n} \left(\overline{Q}_{ij}\right)_k (m_k - m_{k-1}) \\ B_{ij} = \dfrac{1}{2}\sum\limits_{k=1}^{n} \left(\overline{Q}_{ij}\right)_k \left(m_k^2 - m_{k-1}^2\right) \\ D_{ij} = \dfrac{1}{3}\sum\limits_{k=1}^{n} \left(\overline{Q}_{ij}\right)_k \left(m_k^3 - m_{k-1}^3\right) \end{cases} \tag{9}$$

where n is the number of laminate layers on either side of the UD layer (apart from the UD layer) and m_k is the distance between the neutral surface and the outer side of layer k (see Figure 2b) which can be further described by:

$$m_k = kt_f + \frac{t_{UD}}{2} \tag{10}$$

in which t_{UD} and t_f are the thicknesses of the UD layer and the fiber layer (commonly, the thicknesses of the fiber layers using one boom structure were the same).

2.2. Boom Bending Property Calculation

For the sake of illustration, the diagram of a tape spring boom (transversal) cross-section configuration is shown in Figure 4. According to the deployment mode of the boom–membrane structures, the bending torque acted on the boom (after deployment) was mainly around the x or y-axis shown in Figure 4. Hence, the expressions of the bending stiffness around the x and y-axes would be, respectively, studied as follows. Note that O is the center of the boom geometric circle while O_n-x_n is the neutral axis of the boom cross-section configuration in Figure 4.

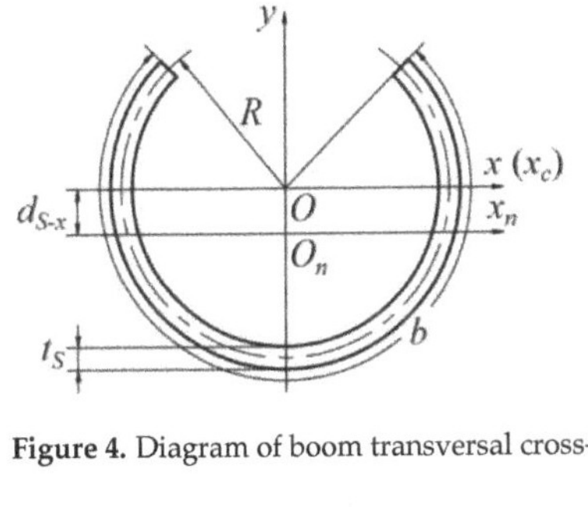

Figure 4. Diagram of boom transversal cross-section configuration.

Based on the geometric configuration shown in Figure 4, the boom's moment of inertia around the x_c-axis can be seen to be:

$$I_{x_c} = 2\int_{R-\frac{t_S}{2}}^{R+\frac{t_S}{2}} R_i dR_i \int_0^{\frac{b}{2R}} R_i^2 \sin^2\theta d\theta \tag{11}$$

where R_i is the variable of integration along radius R, t_S is the total thickness of the boom structure, and θ is the angular variable of integration under a polar coordinate.

On the basis of the parallel-axis formula, the moment of inertia around the x_n-axis is:

$$I_{x_n} = I_{x_c} - d_{S-x}^2 H_S \tag{12}$$

in which H_s is the total area of the boom cross-section configuration and $d_{S\text{-}x}$ showed the distance between x_c and x_n-axes, which can be acquired, respectively, through the equations as follows:

$$H_S = 2\int_{R-\frac{t_S}{2}}^{R+\frac{t_S}{2}} R_i dR_i \int_0^{\frac{b}{2R}} d\phi \tag{13}$$

$$d_{S-x} = \frac{2\int_{R-\frac{t_S}{2}}^{R+\frac{t_S}{2}} R_i dR_i \int_0^{\frac{b}{2R}} R_i \sin\phi d\phi}{H_S} \tag{14}$$

Substituting Equations (13) and (14) into Equation (12) and combining Equations (8) and (12), the boom's bending stiffness around the x-axis can be finally acquired as:

$$S_x = \frac{A_{11} I_{x_n}}{t_S} \tag{15}$$

Similarly, from the diagram in Figure 4, the moment of inertia and the bending stiffness around the y-axis (the y-axis is just the boom's neutral axis in this bending mode) were:

$$I_y = 2\int_{R-\frac{t_S}{2}}^{R+\frac{t_S}{2}} R_i dR_i \int_0^{\frac{b}{2R}} R_i^2 \cos^2\phi d\phi \tag{16}$$

$$S_y = \frac{A_{11} I_y}{t_S} \tag{17}$$

2.3. *Boom Buckling Load Calculation*

For a boom–membrane structure working on-orbit, the boom's buckling failure usually results from a normal tip load caused by the tension requirement of the membrane. According to Equations (15) and (16) and the Euler buckling formula, the critical buckling load of the tape spring boom is:

$$P_{cr} = \frac{\pi^2}{4l^2} \cdot \min\{S_x, S_y\} \tag{18}$$

where l is the total length of the boom. Note that the boundary conditions for the boom buckling calculation is as follows: the boom root is fully fixed with total a free tip (i.e., one end fixed and another end free for each boom), and the load acted (pressed) along the boom length.

2.4. Fundamental Frequency Calculation of Boom–Membrane Structures

On the basis of Refs. [18,19], the fundamental frequency of the boom–membrane structure can be obtained through the formula as shown:

$$f = \frac{1}{2\pi}\sqrt{\frac{n_b S}{l^3(0.2235 \cdot (p_m q_m \rho_{ma} + n_b l H_s \rho_b) + l_r \rho_{rl})}} \tag{19}$$

where n_b is the number of the tape spring booms included in a boom–membrane structure, S is each boom's bending stiffness (the stiffness of each boom is set as the same), p_m and q_m are the length and the width of the membrane, respectively, ρ_{ma} and ρ_b are the densities of the membrane and the tape spring boom (on average), H_s is the area of one boom's cross-section (the same with that used in Equation (12)), l_r is the length of the tip beam, and ρ_{ri} is the linear density of the tip beam.

Note that Equation (19) only aims to acquire the frequency of the modal when the structure is swinging around the root beam (which is the weakest stiffness direction), and the slits of the tape spring booms are commonly in the normal direction of the membrane because of structural limits (hence, $S = S_x$ in Equation (19)). A limitation of the analysis above is that the deformation of the booms should be relatively small, and this condition is in accordance with the common working situations of a boom (used in a membrane structure) working in space.

3. Parametric Study of Tape Spring Booms and Boom–Membrane Structures

To provide more guidance on the boom–membrane structure design, a parametric study is carried out in this section. Table 1 presents the geometric and material parameters of an FPR composite boom which was used for mimicking those used for InflateSail launched in the year 2015 [22,23], while Table 2 lists the parameters of the corresponding boom–membrane structure made of the booms with the parameters shown in Table 1.

Table 1. Geometric and material parameters of tape spring booms.

R (mm)	b (mm)	E_m (GPa)	G_m (GPa)	v_m	E_f (GPa)
20	110	4	2.7	0.35	240
G_f (GPa)	v_f	t_{UD} (mm)	V_{UD} (%)	ϕ_{UD} (%)	t_f (mm)
95	0.22	0.057	31	15	0.096
V_f (%)	ϕ_f (%)	μ	E_h (GPa)	E_r (GPa)	Lay out
53	15	0.1	205	205	$[\pm 50°F/0°]_s$ *

* Mark "F" meant fabric laminate layout.

Table 2. Parameters of boom–membrane structure.

Membrane Size $p_m \times q_m$	Membrane Areal Density ρ_{ma}	Boom Length l	Boom Distance q_m	Boom Density (On Average) ρ_b	Tip Beam Length l_r	Tip Beam Linear Density ρ_{rl}
2950×900 mm^2	0.3 kg/m^2	3000 mm	1000 mm	1500 kg/m^3	1100 mm	0.1 kg/m

Generally, for a boom–membrane structure working in space, the tape spring boom's critical buckling load and the membrane structure's fundamental frequency had great influences on the system's performance. Meanwhile, the effect of a tape spring boom's parameters is sensitive and complicated. Hence, the parametric study of a boom's geometric and material parameters is presented as follows. Note that the parameters of the two booms in a boom–membrane structure can be changed simultaneously in the analysis in this section.

3.1. *Study of Boom Geometric Parameters*

Figure 5 presents the variations in the boom critical buckling load and the membrane structure fundamental frequency under the changing of the boom's cross-section radius R (with a constant path length b) and the total boom length l, respectively. Note that the left vertical axis in each sub-figure is for buckling load (N) while the right axis presents system frequency (Hz), and the parameters that are not marked in the figures are the same as those listed in Tables 1 and 2.

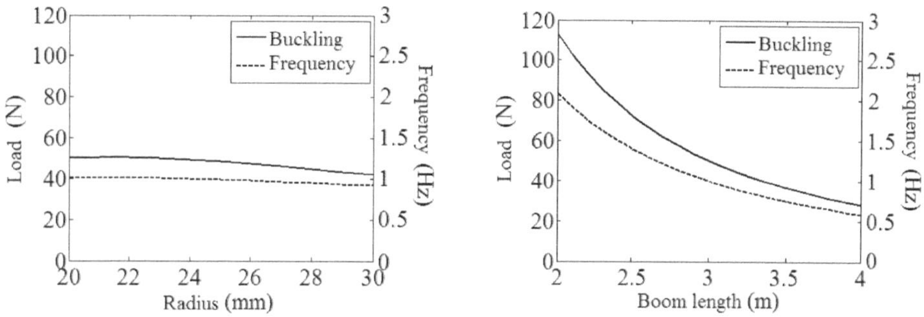

(**a**) Cross-section radius analysis (constant path length) (**b**) Boom length analysis

Figure 5. Boom geometric parametric study.

On the basis of the plots in Figure 5a, it can be observed that both the buckling load and the fundamental frequency decreased with the increase in the boom radius. Meanwhile, generally, these two indicators are less sensitive to the radius change (the boom's bending stiffness changes with the changing of the boom's cross-section radius and the buckling load and the fundamental frequency vary with the change in the bending stiffness). This is because higher R (with a constant path length b) made the boom cross-section flatter, which made the boom more vulnerable to bending torques, and at the same time, the increase in R made the boom's cross-section more scattered, which could partly compensate for the effect mentioned above. From Figure 5b, the load and frequency plots also descend with a growing boom length, and this trend can be relatively more sensitive when the boom is relatively short. This performance is mainly caused by the increase in total mass and the moving outwards of the structure barycenter.

3.2. *Study of Boom Laminate Parameters*

In Figure 6, the influence of the boom laminate parameters, such as the fiber braided angle, the fiber stiffness, the boom wall thickness (overall), and the boom material density (on average), on the boom's buckling load and the membrane structure's fundamental frequency is shown. Note that in Figure 6a, only the braided angle of the fiber layers is changed (every lamina has the same absolute angle value during the analysis) and the angle of the UD layer is always kept constant at $0°$. Moreover, as the density has no connection with the boom's buckling load, only the frequency plot is shown in Figure 6d.

According to the results listed in Figure 6a,b, the buckling load and the fundamental frequency increase with the decreasing of the braided angle or the increase in the fiber stiffness. This is due to the parametric change above enhanced the structural strength and the stiffness along the boom length (direction 1 in Figure 2a). For the plots in Figure 6c, the volume fractions of each lamina were kept constant under the change in the boom's total wall thickness. From Figure 6c, the rise in the wall thickness enhances the tape spring boom's stiffness (this effect also improves the boom's critical buckling load), yet adds the boom's total mass at the same time as well. Therefore, the frequency increases slowly with a growing wall thickness in the plot under these combined actions. Further, it can be seen in Figure 6d that the variation in the boom material density has little effect on the membrane structure's frequency. Therefore, from the result in Figure 6d, reducing the boom density is not an effective way to improve the dynamic stiffness of the boom–membrane structure.

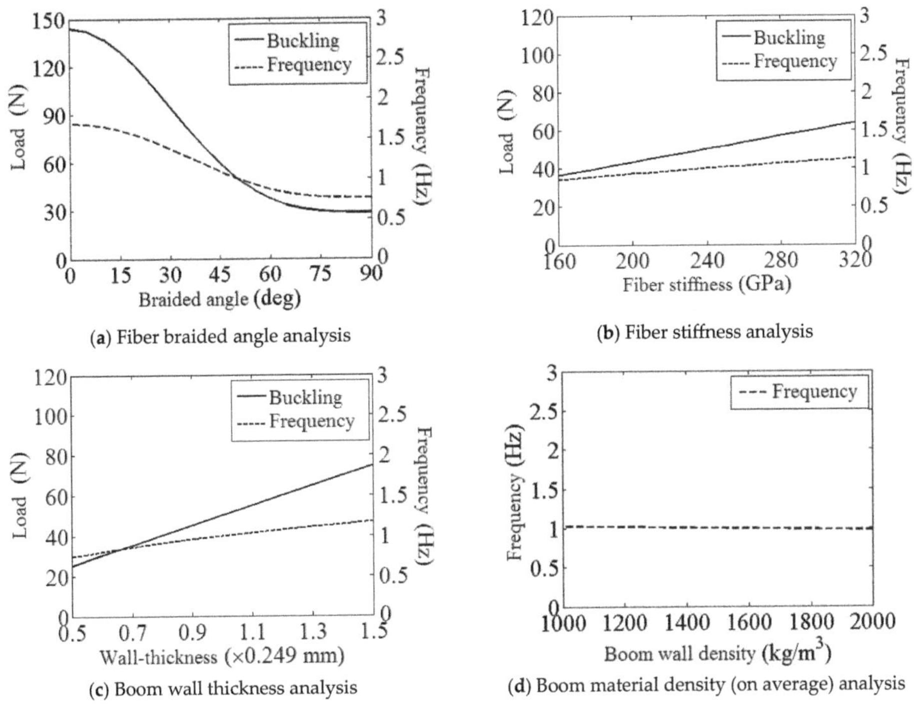

(**a**) Fiber braided angle analysis (**b**) Fiber stiffness analysis (**c**) Boom wall thickness analysis (**d**) Boom material density (on average) analysis

Figure 6. Boom laminate parametric study.

4. Experimental Study

For the sake of providing some practical verification to the theoretical analysis in the above sections, the experiment of the tape spring boom bending stiffness test and the boom–membrane structure frequency test were carried out. Since the boom buckling test is a kind of damage experiment and the critical load directly depends on the boom's bending stiffness according to the Euler formula (Equation (18)), the boom buckling experiment is not necessary for verifying the research in this paper.

4.1. Boom Bending Stiffness Experiment

The parameters of the tape spring boom sample used for the experiment were R = 25 mm, b = 137.5 mm, l = 2000 mm and the boom's laminate layout was set as $[\pm 50°F/0°]_S$ (carbon-fiber-reinforced polymers) in order to better manufacture (the other parameters were the same with those listed in Table 1). For keeping the configurations of the tip and the root of the boom sample during the bending test process, customized plugs were introduced at the tip and the root of each sample (see Figure 7a).

Figure 7b shows the facilities used for the single-boom bending stiffness experiment. In the figure, the boom root was fully fixed (all DOFs restrained) while the boom tip was hanging up on the ceiling with a string for compensating the boom and the plug gravity. The tip was loaded by weights through a fixed pulley, and the tip displacement was acquired and recorded by a laser displacement sensor which was fixed on the test bed (fixed on the ground). The value of the counterweight was selected on the premise of small deformation of the boom. Further, the boom tip displacement and the counterweight was converted into bending stiffness (S_e) through the following equations (see Figure 8 for the corresponding conversion diagram):

$$\begin{cases} (r_e - d_e)^2 + \left(r_e \cdot \sin\left(\frac{l}{r_e}\right)\right)^2 = r_e^2 \\ S_e = M_e \cdot r_e \end{cases} \tag{20}$$

where M_e is the bending torque acted on the boom tip caused by the weights (see Figure 8 for other symbol definitions in Equation (20)). Before the beginning of the experiment, the laser displacement sensor was calibrated and reset as zero for the measurements, and the results obtained from the laser sensor after the experiment can be combined with the counterweight values based on the conversion in Figure 8 and Equation (20).

(**a**) Boom plug

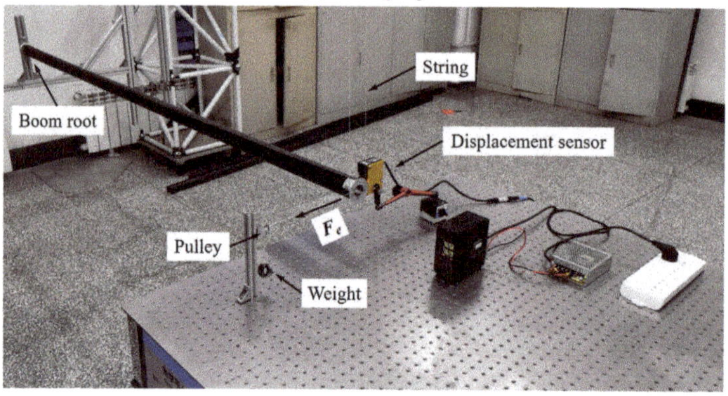

(**b**) Boom bending stiffness experimental facility

Figure 7. Boom bending stiffness experiment.

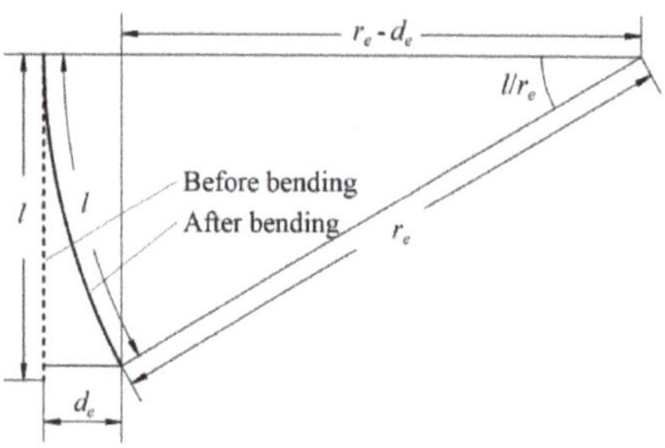

Figure 8. Diagram of boom bending stiffness conversion.

The boom bending stiffness experiment results are listed in Table 3, and the stiffness around x and y-axes were listed and compared with the corresponding theoretical results, respectively. According to the comparison, it can be observed that the testing and analytical results matched with each other very well, and all the experimental relative errors were below 2%. Meanwhile, a higher counterweight led to lower testing errors as appropriate weight made the boom bending more ideal. To sum up, from the testing experiment it can be known that the analytical method in this paper is available for solving the bending stiffness and the buckling load of an FRP composite laminate tape spring boom.

Table 3. Experiment results of boom bending stiffness.

	Weight (g) Corresponding torque (N·m)		5 0.98	10 1.96	15 2.94	20 3.92
Around x-axis	Displacement (mm)		11.46	22.92	34.39	45.88
	Bending stiffness	Test (N·m^4)	349.0	349.0	348.9	348.7
		Theory (N·m^4)	343.4			
		Error (%)	−1.63	−1.63	−1.60	−1.54
Around y-axis	Displacement (mm)		8.12	16.29	24.56	32.98
	Bending stiffness	Test (N·m^4)	492.6	491.1	488.6	485.1
		Theory (N·m^4)	484.3			
		Error (%)	−1.71	−1.40	−0.98	−0.17

4.2. Fundamental Frequency Experiment of Boom–Membrane Structure

In the fundamental frequency experiment, the geometric parameters of the boom–membrane structure prototype were changed to the values presented in Table 4 for the ease of manufacturing (the other parameters were in accordance with those values in Table 2). Based on the corresponding parameters listed in Table 4, the boom–membrane structure prototype is made and shown in Figure 9a. Figure 9b presents the fundamental experimental facilities. In the figures, the roots of the two tape spring booms were fixed on the test bed (all DOFs restrained). The boom tips were connected with a tip beam (plugs introduced on each boom's tip and root as shown in Figure 7a and the weight of each plug was measured as 0.5 kg), and the membrane (made from polyimide) was tensioned between the tip beam and the root beam close to the test bed. The membrane was tensioned between the root and tip beams, while the acceleration sensors were assigned and attached along the tape spring booms sending the testing data to the LMS vibration testing system (developed by Siemens). Through the LMS Scadas III Test Lab resolving software and interface, the vibration frequency of the boom–membrane structural prototype can be displayed on the computer. The hammer point was selected at the middle of the tip beam based on the previous testing experience. To improve the reliability of the experimental results, four tests were carried out during the frequency experiment. Before the fundamental frequency experiment, the LMS vibration testing system was calibrated by testing a steel plain with known vibration properties. During the test, the boom–membrane structural prototype stood on the test bed (which was fully fixed on the ground) like a vertical cantilever with its two boom roots and one root beam wholly restrained on the test bed (see Figure 9), which was used for mimicking the zero-gravity condition when working on-orbit. The damping factor was ignored for the experiment since the oscillating velocity was relatively low; hence, the air-damping and the structural damping were negligible.

Table 4. Parameters of boom–membrane structure used for experimental study.

Membrane size $p_m \times q_m$ 1980 × 450 mm^2	Membrane areal density ρ_{ma} 0.15 kg/m^2	Boom length l 2000 mm	Boom distance q_m 570 mm
Boom density (on average) ρ_b 1500 kg/m^3	Tip beam length l_r 620 mm	Tip beam linear density ρ_{rl} 0.05 kg/m	Membrane tension T_m 10 N

(**a**) Membrane structure prototype

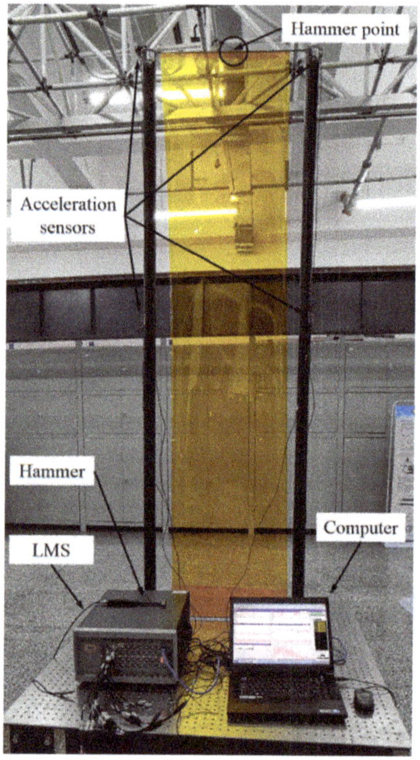

(**b**) Fundamental frequency testing system

Figure 9. Membrane structure prototype and fundamental frequency experimental facility.

The experimental results of the membrane structure's fundamental frequency are shown in Table 5, and the corresponding theoretical results were also listed for comparison. On the basis of the results in Table 5, it can be seen that the experimental results generally matched with the theory, and the errors of the four tests were all below 10%. Furthermore, from Table 5, experimental results were slightly higher than the theory. This was because the structure stood on the test bed; hence, the gravity component would reduce the system's vibration frequency. Nevertheless, this experiment could verify that the theory in this paper is available for predicting the fundamental frequency of this kind of boom–membrane structure. By comparing with the investigation in Refs. [24,25], the method in this paper better balanced the model complexity and the result accuracy for acquiring the properties of the FRP composite booms and the corresponding boom–membrane structures.

Table 5. Experimental results of boom–membrane structure fundamental frequency.

	Test 1	Test 2	Test 3	Test 4
Experiment (Hz)	3.08	3.26	3.10	3.15
Theory (Hz)	3.32			
Error (%)	+7.8	+1.8	+7.1	+5.4

5. Conclusions and Discussion

Extendable deployable tape spring boom mechanisms were commonly used for extra large-scale space structures which were usually combined with flexible membranes forming planar boom–membrane structures, which can meet the needs of large planar space structures. The early tape spring booms were usually made from metal materials, while FRP laminate composites have often been used to improve the stability and design flexibility of the booms recently. However, the research of the previous work is mainly concentrated on the boom deploying process, and the properties of the deployed and deploying states were usually studied independently.

The work of this paper aimed to analyze the deployed state of the FRP composite laminate tape spring booms and the corresponding boom–membrane structures. For this purpose, the analytical models of the laminate tape spring boom and the boom–membrane structure were established to solve the boom's bending stiffness, buckling loads and the membrane structure's fundamental frequency. For the sake of understanding the influence of the key design parameters, a parametric study was carried out to afford more guidance for a better design. Based on the geometric parameter analysis, it can be found that the buckling and frequency properties are relatively insensitive to the boom's cross-section radius change (with a constant path length) while they are sensitive to the boom's total length. On the basis of the laminate parametric study, both of the indicators grow with the enhancement of the boom bending stiffness and the boom wall thickness.The boom material density has a slight effect on the system's fundamental frequency. To provide some practical verification, the experiment on a tape spring boom and the boom–membrane structure was carried out. Comparing the stiffness and frequency results from the experiment and the theory, it can be observed that the analytical method in this paper is available for predicting the deployed properties of the booms and the structures.

Furthermore, in the design of deployable structure mechanisms, the properties of the structure in both deploying and deployed states should be considered comprehensively. The work in this paper follows up on the research of the authors' previous study which aimed at the boom–membrane structure deploying process. The multi-configuration optimization of the structure will be further researched and published in the near further, which will be on the basis of the work in this paper.

Author Contributions: Conceptualization, S.X. and X.Y.; methodology, X.Y. and Y.G.; formal analysis, S.X.; writing—review and editing and investigation, X.Y.; resources, Y.G.; writing—original draft and funding acquisition, S.W.; supervision and visualization, L.S. All authors have read and agreed to the published version of the manuscript.

Funding: The research is funded by the National Natural Science Foundation of China (Grant No. 52205027) and the Science and Technology Project of Jiangsu Province (Grant No. BK20220496).

Institutional Review Board Statement: Not applicable.

Informed Consent Statement: Not applicable.

Data Availability Statement: The data presented in this study are available on request from the corresponding author. The data are not publicly available due to relevant policy restrictions.

Acknowledgments: The authors would like to acknowledge that the research presented in this paper was carried out with the aid of the National Natural Science Foundation of China (Grant No. 52205027) and the Science and Technology Project of Jiangsu Province (Grant No. BK20220496).

Conflicts of Interest: The authors declare no conflicts of interest.

References

1. Sim, A.; Santer, M. Analysis of a segmented dompliant deployable boom for CubeSat magnetometer missions. In Proceedings of the 51st AIAA/ASME/ASCE/AHS/ASC Structures, Structural Dynamics, and Materials Conference, Orlando, FL, USA, 12–15 April 2010; p. 2581.
2. Leclerc, C.; Pedivellano, A.; Pellegrino, S. Stress concentration and material failure during doiling of ultra-thin TRAC booms. In Proceedings of the 2018 AIAA Spacecraft Structures Conference, Kissimmee, FL, USA, 8–12 January 2018; p. 690.
3. Bourgeois-Doyle, R.; Klein, G. *The Great Inventor: NRC Press Biography Series*; NRC Research Press: Ottawa, ON, Canada, 2004; No. 2.
4. Suliga, A.; Jakubczyk, E.; Hamerton, I.; Viquerat, A. Analysis of atomic oxygen and ultraviolet exposure effects on cycloaliphatic epoxy resins reinforced with octa-functional POSS. *Acta Astronaut.* **2018**, *142*, 103–111. [CrossRef]

5. Cox, K.; Davis, B.; VanHalle, R.; Francis, W. Flight build of a furled high strain composite antenna for CubeSats. In Proceedings of the 2018 AIAA Spacecraft Structures Conference, Kissimmee, FL, USA, 8–12 January 2018; p. 1678.
6. Martin, E.; Hassan, R. Comparing the N-branch genetic algorithm and the multi-objective genetic algorithm. *AIAA J.* **2004**, *42*, 1496–1500. [CrossRef]
7. Santer, M.; Pellegrino, S. An asymmetrically-bistable monolithic energy-storing structure. In Proceedings of the 45th AIAA/ASME/ASCE/AHS/ASC Structures, Structural Dynamics & Materials Conference, Palm Springs, CA, USA, 19–22 April 2004; p. 1527.
8. Zhang, Z.; Yang, L.; Yu, X.; Li, X.; Wu, H.; Wu, H.; Jiang, S.; Chai, G. Bistable morphing composite structures: A review. *Thin-Walled Struct.* **2019**, *142*, 74–97. [CrossRef]
9. Pellegrino, S. Bistable Shell Structures. In Proceedings of the 46th AIAA/ASME/ASCE/AHS/ASC Structures, Structural Dynamics and Materials Conference, Austin, TX, USA, 18–21 April 2005; p. 1934.
10. Sickinger, C.; Herbeck, L.; Breitbach, E. Structural engineering on deployable CFRP booms for a solar propelled sailcraft. In Proceedings of the 54th International Astronautical Congress of the International Astronautical Federation, the International Academy of Astronautics, and the International Institute of Space Law, Bremen, Germany, 29 September–3 October 2003; Volume 58, pp. 185–196.
11. Murphy, D. MegaFlex—The scaling potential of UltraFlex technology. In Proceedings of the 53rd AIAA/ASME/ASCE/AHS/ASC Structures, Structural Dynamics and Materials Conference, Honolulu, HI, USA, 23–26 April 2012; p. 1581.
12. Costa, M. A spaceborne small SAR design with reflector antenna associated with compact polarimetry. In Proceedings of the AIAA SPACE, Long Beach, CA, USA, 13–16 September 2016; p. 5499.
13. Mallol, P.; Tibert, G. Deployment modeling and experimental testing of a bi-stable composite boom for small satellites. In Proceedings of the 54th AIAA/ASME/ASCE/AHS/ASC Structures, Structural Dynamics, and Materials Conference, Boston, CA, USA, 8–11 April 2013; p. 1672.
14. Mao, H.; Sinn, T.; Vasile, M.; Tibert, G. Simulation and control of a space web deployed by centrifugal forces in a sounding rocket experiment. In Proceedings of the AIAA Modeling and Simulation Technologies Conference, Washington, DC, USA, 13–17 June 2016; p. 4421.
15. Santer, M.; Sim, A.; Stafford, J. Testing of a segmented compliant deployable boom for CubeSat magnetometer Missions. In Proceedings of the 52nd AIAA/ASME/ASCE/AHS/ASC Structures, Structural Dynamics and Materials Conference, Denver, CO, USA, 4–7 April 2011; p. 1732.
16. Chu, Z.; Lei, Y. Design theory and dynamic analysis of a deployable boom. *Mech. Mach. Theory* **2014**, *71*, 126–141. [CrossRef]
17. Pellegrino, S. Large retractable appendages in spacecraft. *J. Spacecr. Rocket.* **1995**, *32*, 1006–1014. [CrossRef]
18. Roybal, F.; Banik, J.; Murphey, T. Development of an elastically deployable boom for tensioned planar structures. In Proceedings of the 48th AIAA/ASME/ASCE/AHS/ASC Structures, Structural Dynamics, and Materials Conference, Honolulu, HI, USA, 23–26 April 2007; p. 1838.
19. Banik, J.; Murphey, T. Performance validation of the Triangular Rollable And Collapsible mast. In Proceedings of the 24th Annual AIAA/USU Conference on Small Stallites, Logan, Utah, USA, 9–12 August 2010; p. 2159.
20. Wang, S.; Schenk, M.; Jiang, S.; Viquerat, A. Blossoming analysis of composites deployable booms. *Thin-Walled Struct.* **2020**, *157*, 107098. [CrossRef]
21. Wang, S.; Schenk, M.; Guo, H.; Viquerat, A. Tip force and pressure distribution analysis of a deployable boom during blossoming. *Int. J. Solids Struct.* **2020**, *194*, 141–151. [CrossRef]
22. Underwood, C.; Viquerat, A.; Schenk, M.; Taylor, B.; Massimiani, C.; Duke, R.; Stewart, B.; Fellowes, S.; Bridges, C.; Aglietti, G.; et al. Inflatesail de-orbit flight demonstration results and follow-on drag-sail applications. *Acta Astronaut.* **2019**, *162*, 344–358. [CrossRef]
23. Viquerat, A.; Schenk, M.; Lappas, V.; Sanders, B. Functional and qualification testing of the InflateSail technology demonstrator. In Proceedings of the 2nd AIAA Spacecraft Structures Conference, Kissimmee, FL, USA, 5–9 January 2015; p. 1627.
24. Wu, C.; Viquerat, A. Natural frequency optimization of braided bi-stable carbon/epoxy tubes: Analysis of braid angles and stacking sequences. *Compos. Struct.* **2017**, *159*, 528–537. [CrossRef]
25. Wu, C.; Viquerat, A.; Aglietti, G. Optimization of the natural frequency and buckling analysis of bi-stable carbon fiber reinforced plastic booms for space applications. In Proceedings of the 3rd AIAA Spacecraft Structures Conference, San Diego, CA, USA, 4–8 January 2016; pp. 156–168.

materials

MDPI

Article

Elastic Property Evaluation of Fiberglass and Epoxy Resin Composite Material Using Digital Image Correlation

Dalferson Yoras [1], Sylwia Makowska [2], Agnė Kairytė [3,*], Jurga Šeputytė-Jucikė [3], Luis Roberto Centeno Drehmmer [1] and Maikson Luiz Passaia Tonatto [1]

1 Group on Mechanics of Materials and Structures, Campus Cachoeira do Sul, Federal University of Santa Maria, Cachoeira do Sul 96503-205, Brazil; dalfersonyoras123@hotmail.com (D.Y.); luis.drehmer@ufsm.br (L.R.C.D.); maikson.tonatto@ufsm.br (M.L.P.T.)

2 Institute of Polymer and Dye Technology, Faculty of Chemistry, Lodz University of Technology, Stefanowskiego 12/16, 90-924 Lodz, Poland; sylwia.czlonka@edu.p.lodz.pl

3 Laboratory of Thermal Insulating Materials and Acoustics, Institute of Building Materials, Faculty of Civil Engineering, Vilnius Gediminas Technical University, Linkmenų St. 28, 08217 Vilnius, Lithuania; jurga.seputyte-jucike@vilniustech.lt

* Correspondence: agne.kairyte@vilniustech.lt

Abstract: This study focused on evaluating the sensitivity and limitations of the simplified equipment used in the Digital Image Correlation (DIC) technique, comparing them with the analog extensometer, based on the mechanical property data of a composite made of fiberglass and epoxy resin. The objectives included establishing a methodology based on the literature, fabricating samples through manual lamination, conducting mechanical tests according to the ASTM D3039 and D3518 standards, comparing DIC with the analog extensometer of the testing machine, and contrasting the experimental results with classical laminate theory. Three composite plates with specific stacking sequences ($[0]_3$, $[90]_4$, and $[\pm45]_3$) were fabricated, and samples were extracted for testing to determine tensile strength, modulus of elasticity, and other properties. DIC was used to capture deformation fields during testing. Comparisons between data obtained from the analog extensometer and DIC revealed differences of 11.1% for the longitudinal modulus of elasticity E_1 and 5.6% for E_2. Under low deformation conditions, DIC showed lower efficiency due to equipment limitations. Finally, a theoretical analysis based on classical laminate theory, conducted using a Python script, estimated the longitudinal modulus of elasticity E_x and the shear strength of the $[\pm45]_3$ laminate, highlighting a relative difference of 31.2% between the theoretical value of 7136 MPa and the experimental value of 5208 MPa for E_x.

Keywords: composite materials; strain measurements; epoxy resin; fiberglass; manual lamination; mechanical properties

1. Introduction

Composite materials highlight their significant structural efficiency, demonstrating a strong relationship between high strength and low weight [1,2]. These high-strength polymeric materials have the potential to replace metals and can constitute up to 50% of the total mass of aircraft. The use of composites in aircraft manufacturing offers several advantages, including weight reduction, improved structural efficiency, corrosion resistance, and adaptable design, as well as reduced vibrations and noise, as noted by [3]. Research in composite materials is crucial for advancing innovative technologies and addressing contemporary challenges. Baggio et al. [4] emphasizes the importance of understanding the characteristics and behavior of these materials for their proper configuration in terms of sizing, construction, and application.

Strain measurements are essential for obtaining precision in elastic properties, promoting a deeper understanding of the mechanical behavior of these materials. The heterogeneity of composite materials [5,6] makes the strain measurements, whose properties are influenced not only by the material composition but also by the geometric design of the structural elements, particularly difficult.

From a theoretical perspective, analytical theories can estimate the elastic constants of laminates. Macromechanical behavior refers to the average macroscopic mechanical properties of laminates, which can be analytically evaluated from a theoretical perspective. Researchers, such as Belaid et al. [7], indicate that these properties can be obtained through experimental tests or calculated based on micromechanics. This allows for the estimation of laminate behavior under various load combinations. According to the classical laminate theory, as discussed by Kaw [8], theoretical data on global and local stresses, strains, and potential failures can be obtained. This theory assumes an elastic–linear behavior for composite materials, a characteristic that has been verified in E-glass/epoxy laminates. Neto [9] highlights that composite materials exhibit a greater capability of following a linear stress–strain relationship when subjected to loading compared to metallic materials, making it favorable to consider them as linear elastic materials.

From an experimental perspective, Digital Image Correlation (DIC) techniques are widely employed for this purpose, offering precise analysis of displacements and strains. This capability is essential for characterizing and optimizing composite materials, which are extensively used in industries such as aerospace, military, automotive, and sports. DIC works by capturing the displacement vector of each pixel point on the sample surface through a comparison of its position in images taken before loading and after deformation. The accuracy of this process heavily relies on the presence of a unique and random pattern of markers, as emphasized by Sabik et al. [10].

An NBC News report [11] with experts discussing the implosion of the submersible emphasizes the need for numerous specific tests and analyses for this type of material. The analysis of correlation between the initial and final images allows for determining displacement fields, from which sample deformations are derived. The DIC-2D technique requires only a high-definition camera positioned orthogonally to the surface of the test object, maintaining a fixed distance from the camera to the sample, as emphasized by [12,13]. Calibration is based on comparing a known measurement with the results obtained by the DIC technique, involving the evaluation of different deformation states to increase the reliability of the generated values. Blenkinsopp et al. [14] demonstrated the importance of establishing precise predefined deformations used as the reference.

Recent studies have explored laminates of composite materials, such as carbon fiber reinforced polymers and thermoplastics, widely used in the aerospace industry. Non-destructive approaches, such as DIC, have been employed to examine the behavior of these materials. Some studies investigated the influence of manufacturing defects and stress concentrations in quasi-static fatigue tests [15]. Others analyzed displacements in steel beams using DIC, dividing the work into experimental setup, image processing, and result comparison [16]. Gonabadi et al. [17] studied the effects of anisotropy in composite structures for renewable energy, such as wind turbine blades, using DIC and microscopy analysis. Feito et al. [18] analyzed carbon fiber reinforced polymer (CFRP) laminates using non-destructive techniques like IRT and DIC to characterize manufacturing defects and stress concentrators during fatigue tests. Hao et al. [19] reported that the DIC technique allowed for full-field displacements and deformations, identifying potential failures.

The study of composite materials, particularly in relation to their mechanical behavior under various conditions, has seen extensive research employing advanced techniques such as DIC. Researchers have investigated aspects like fiber orientation, strain localization, static and fatigue properties, failure modes, and strain distribution using DIC [20,21], thermographic stress analysis [22], Acoustic Emission [23], and conventional mechanical testing [24,25]. This array of studies underscores the versatility and importance of DIC in providing detailed insights into the mechanical behavior of fiber-reinforced composites. However, the sensitivity and limitations of lower-cost equipment in DIC, especially for measuring small strains, are not yet well understood. Identifying the detection limits of strain and displacements for DIC and analog extensometers is crucial for further development in this field.

This study focused on exploring the sensitivity and limitations of DIC techniques and analog extensometers in assessing the elastic properties of fiberglass/epoxy laminates, especially in small strains.

2. Materials and Methods

With the aim of developing an assessment of the sensitivity and limits of the equipment used in the DIC technique, the methodology employed for constructing the composite was executed in a standardized manner to minimize potential variations among samples. For the three produced laminated plates, both the fiber and matrix were used in equal proportions. Parameters such as temperature, compaction weight, and curing time were kept constant. Ten samples were taken from each plate, with only the best five samples from each laminate considered for statistical data analysis.

2.1. Materials and Manufacturing

The manual lamination process and mechanical tests were conducted in the Mechanical Engineering Laboratory of the Federal University of Santa Maria—Campus Cachoeira do Sul.

Figure 1 presents a flowchart describing the main steps that were followed during the conduct of this study, which was planned to be carried out systematically.

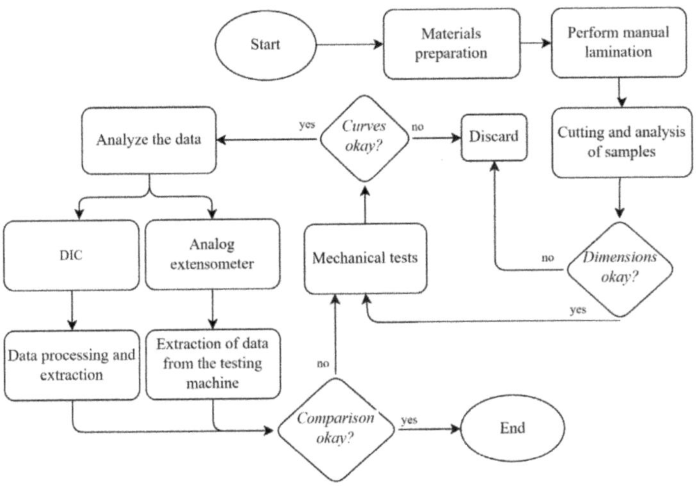

Figure 1. Methodology flowchart.

In the lamination process, uni-directional E-glass fiber fabric from Texiglass [26] was used in conjunction with fast-curing, low-resistance epoxy resin AR-320, combined with hardener AH-320 from the E-composites brand (E-Composites, Rio de Janeiro, Brasil). To ensure proper compaction of the composite and minimize the formation of air bubbles in the laminate, nylon peel-ply fabric was employed. Additionally, a release agent such as carnauba wax was applied to facilitate the removal of the composite plates after the curing process, along with other materials used in manual lamination.

During the process, three composite plates were fabricated, each with specific dimensions due to variations in sample sizes. Each plate was sized to accommodate a total of ten samples, including tabs. The plates were produced with varying numbers of layers and fiber orientations to meet thickness specifications required by applicable standards for composite materials testing. In the $[0]_3$ configuration, three layers of fabric were aligned in the same direction as the applied tensile load. In the $[90]_4$ configuration, four layers of fabric were positioned perpendicular to the direction of the tensile load. Finally, in the $[\pm45]_3$ configuration, a total of six layers of fabric were interleaved.

The total curing time for the plates was 36 h, with a controlled temperature of 22 °C in the laboratory, and a compaction weight of 35 kg applied to each plate. Cutting of the plates for sample and tab preparation was conducted according to dimensions specified in standards and detailed in Table 1. A computer numerical control (CNC) machine was utilized for the cutting process, employing water for material cooling to prevent tool clogging.

Table 1. Table of sample sizing.

Stacking Sequence	Number of Samples	Dimensions of the Samples	Dimensions of the Plate
$[0]_3$	10	[1] 250 × 15 mm	315 × 230 mm
$[90]_4$	10	[1] 175 × 25 mm	250 × 310 mm
$[\pm45]_3$	10	[1] 250 × 25 mm	275 × 280 mm

[1] Dimensions in (length × width).

For the tab bonding process, the same resin used in composite fabrication was employed, necessary only in the $[0]_3$ configuration to prevent slippage during testing and reduce sample failures in the machine grip region.

Additionally, a final step was required for the samples, which involved surface preparation to enable effective image capture using the DIC technique. This included applying a uniform layer of white paint on all samples, followed by a second coat of black paint applied non-uniformly to create small splatters using a specific and deliberate technique, as shown in Figure 2. This surface preparation was crucial to facilitate the use of the DIC method.

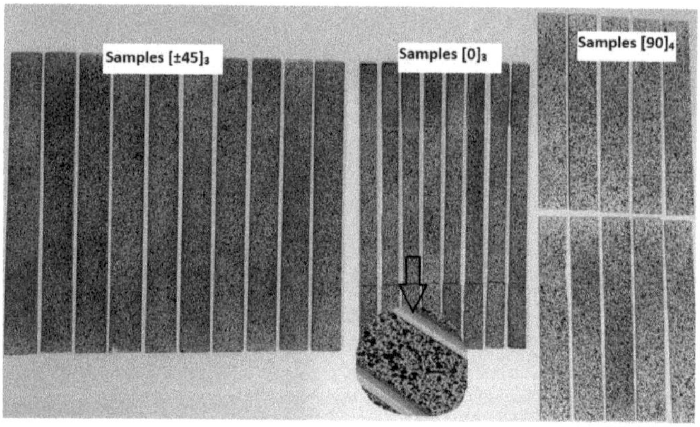

Figure 2. Samples with surface preparation.

2.2. Mechanical Tests

The mechanical tests were conducted following the guidelines established by ASTM D3039 [27] for composite material tensile testing. In the $[0]_3$ configuration, stress–strain curves were obtained to determine the modulus of elasticity (E_1), longitudinal tensile strength (σ_1^T), and (ν_{12}). Additionally, tests in the $[90]_4$ configuration aimed to determine the modulus of elasticity (E_2) and longitudinal tensile strength (σ_2^T).

Shear testing according to ASTM D3518 [28] was performed on $[\pm45]_3$ samples, allowing for the determination of shear modulus (G_{12}), shear strength (τ_{12}), and the elastic modulus (E_x) of the laminate.

The tests were conducted on a set of thirty samples, evenly distributed with ten samples for each configuration. This procedure resulted in a total of sixty distinct analyses, as each specimen was evaluated using two different methods. Both standards require a minimum of five samples, leading to the selection of the top five performing samples for each configuration. Consequently, the average mechanical properties were calculated exclusively based on these selected five samples.

The evaluations were carried out using the LEIDA brand universal testing machine, model LW-5000 (Leida equipamentos para laboratórios e ensaios mecânicos, São Paulo, Brasil), with a maximum load capacity of 500 kN, resolution of 0.01 kN, and an approximate error margin of ±1%. The test speed used was 2 (mm/min), as specified by the standard. Figure 3a,b illustrate the configuration of the devices used during the mechanical tests.

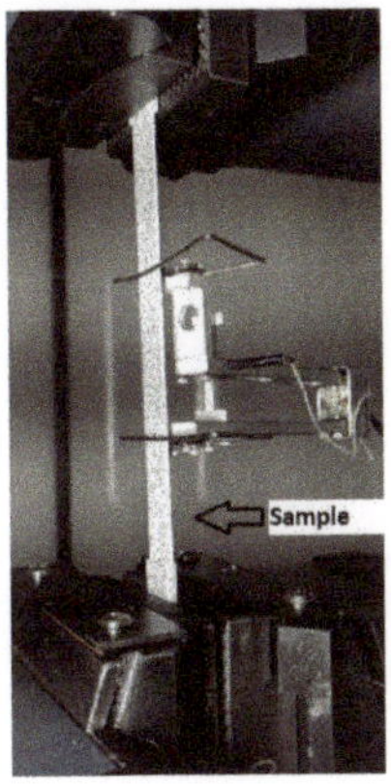

(**a**)

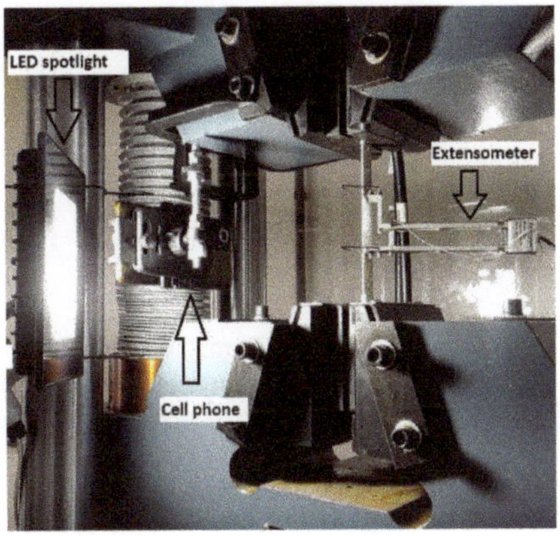

(**b**)

Figure 3. (**a**) Sample clamped in the machine setup. (**b**) During the mechanical test.

The analog extensometer was initially used to record strain data during mechanical tests, as illustrated in Figure 3a. The calculation of composite properties began with determining the modulus of elasticity (E) in direction i, obtained by the difference in tensile stress ($\Delta\sigma$) divided by the difference in strain ($\Delta\epsilon$), following the standard of 1000 $\mu\epsilon$ to 3000 $\mu\epsilon$. The shear modulus was calculated similarly, with deformations from 1500 $\mu\epsilon$ to 4000 $\mu\epsilon$.

The tensile strength of the laminates was recorded by the testing machine. The Poisson's ratio was calculated by the variation in transverse strains relative to longitudinal strains, considering the same initial and final points used in the calculation of the modulus of elasticity.

The dimensions of the test specimens were measured using a 150 mm digital caliper with a resolution of 0.01 mm, following the normative recommendation to measure at least three different points on the sample. During the mechanical tests, images were captured using a strategically positioned smartphone to ensure the best field of view, aided by an LED light for adequate illumination under the sample, as shown in Figure 3b.

Videos were recorded during the tests using a Xiaomi smartphone, model Redmi Note 10 Pro (Xiaomi, Beijing, China), equipped with a 108-megapixel rear camera. This camera was capable of recording videos in 4K resolution, achieving up to 3840 × 2160 pixels and a frame rate of 30 frames per second (FPS). Subsequently, the videos were converted into individual images at 1 s intervals and then imported for analysis into the image processing software to obtain deformation fields. The GOM Correlate 2022 software from Aramis was used for this analysis.

The GOM Correlate software [29] uses the DIC technique. The technique is used to measure the displacements and strains on surfaces by comparing high-resolution images captured before and after deformation. It operates by dividing these images into subsets, identifying unique patterns within each subset, and correlating these patterns between the two images to compute local displacements. Using methods like normalized cross-correlation, DIC calculates displacement vectors for each subset, which are then interpolated to generate full-field displacement maps. From these maps, strains can be derived to analyze the mechanical behavior of materials. This software transforms images into deformation data, allowing comparisons with the analog extensometer. Consequently, it was possible to compile the data into spreadsheets, providing a comprehensive statistical analysis of all tested samples.

The classical laminate theory was used to calculate the engineering constant E_x of $[\pm45]_3$ laminate. A script was used in order to automate the resolution of equations, and calculation of stresses, strains, elastic modulus, and failure estimates of the laminate, both on a global and local scale. The mechanical properties for $[0]_3$ and $[90]_4$ laminates obtained from tests were used as input. This enabled a comparison between the experimental results of the mechanical tests and the theoretical predictions of classical laminate theory for the $[\pm45]_3$ laminate, using global coordinates due to the fiber orientation.

3. Results and Discussion

In the initial phase, the results of the $[0]_3$, $[90]_4$, and $[\pm45]_3$ configurations were presented and discussed using figures, tables, and comparative graphs of the samples. Subsequently, the obtained results from the three tests were analyzed in relation to the relevant literature.

Following the completion of individual tests, a detailed approach was adopted. All collected data were migrated to Excel software, necessitating a meticulous alignment process to correlate the results from the two analysis methods.

An important consideration was the lack of synchronization between the recorded video during the tests and the actual start time of the mechanical test on the machine. To address this, the video recording began a few seconds before the effective start of the mechanical test. This initial time lag received careful attention during the Excel analysis. The purpose of this care was to ensure precise and reliable comparisons of data obtained through these complementary methods, thereby maintaining the integrity and relevance of the final results.

The limit in capturing deformations using the DIC technique with simplified equipment was explored in our study. Despite the challenges posed by simplified setups, valuable insights were gained into the material's behavior under deformation. A thorough analysis of these limits not only outlined the boundaries for reliable measurements but also underscored the importance of understanding equipment constraints in DIC applications. This understanding enabled a precise interpretation of the experimental results, ensuring that the conclusions were firmly grounded in the capabilities of the technology employed.

The results section is subdivided for each test configuration. The first series covers the $[0]_3$ configuration, followed by the $[90]_4$ configuration, and finally the $[\pm45]_3$ configuration. These divisions were necessary due to specific adjustments required on the testing machine for each configuration.

3.1. *Results of Laminate* $[0]_3$

Initially, greater mechanical strength was observed in the tests because the fibers were aligned with the applied load. However, these specimens also showed low strains, which hindered detailed analysis using the DIC method, largely due to the limited resolution of the equipment used.

An important consideration was made regarding the differences between the methods employed. The primary focus of the tests was on the elastic regime. Therefore, the analog extensometer was removed after reaching approximately 0.7% of strain.

Additionally, during the tests, strains were captured using the DIC method. Since this approach does not require direct contact of the equipment with the sample, it allows for a detailed evaluation of the strain field and the sample's sensitivity, complementing the information obtained by the analog method. This expands the range of the elastic constants obtained, such as the Poisson's ratio, for which it is necessary to use two extensometers simultaneously during the test.

Figure 4a displays the strain field along sample 2, representing configuration $[0]_3$. The strain field in the fiber direction is observed at a load level of 13.7 kN using DIC. In these representations, digital extensometers are simulated on each specimen to extract strain data.

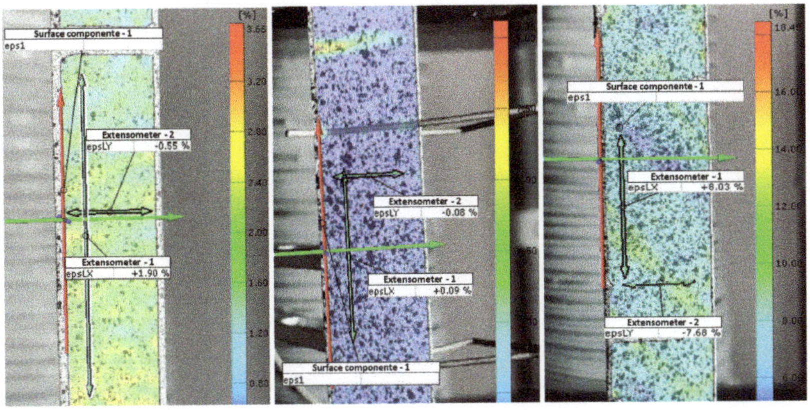

(**a**) (**b**) (**c**)

Figure 4. Scheme of longitudinal and transverse digital extensometers for the configurations: (**a**) $[0]_3$, (**b**) $[90]_4$, and (**c**) $[\pm 45]_3$.

In the graph in Figure 5, through the stress–strain diagram, ten curves are presented, five of which were captured by the analog extensometer and the other five by the digital extensometer (DIC). Only the strain range required by the standard was demonstrated. A variation between the curves was evident when comparing the strains from the two procedures; however, good agreement was observed, and a trend could be discerned when comparing all curves from the same method.

Based on a statistical study, the five best curves were selected and, from these, the mean, standard deviation, and coefficient of variation were calculated to characterize this laminate. Despite the process variability and experimental measurement, the larger number of samples tested and the selection of the best ones reduced the uncertainties in the results.

It was found that the curves obtained with the analog extensometer exhibited greater uniformity than those obtained with DIC. This could be explained by two reasons: first, the analog method collected data at intervals of 0.1 s, while the digital method did so every 1 s, resulting in fewer points to plot the graph, affecting the linearity of the curves. The second point to consider is that this configuration involved low strains, making it difficult for the camera to capture the deformation in the same proportion as the analog extensometer, showing a certain limitation of the equipment, such as the camera. In other words, the DIC captured, for example, a variation of about 0.3% strain in a shorter time interval than the analog extensometer, showing that it has lower sensitivity to low deformations. Therefore, when comparing the curves for the same specimen at the same strain point, the DIC curve presented lower stress.

Table 2 presents the results of the elastic moduli, one obtained through the analog extensometer and the other with the digital extensometer using the DIC technique. Additionally, the tensile strength values in the fiber direction were described, obtained through the load cell of the testing machine, relating the applied force to the cross-sectional area of the sample. Furthermore, Poisson's ratio in the 12-plane was included.

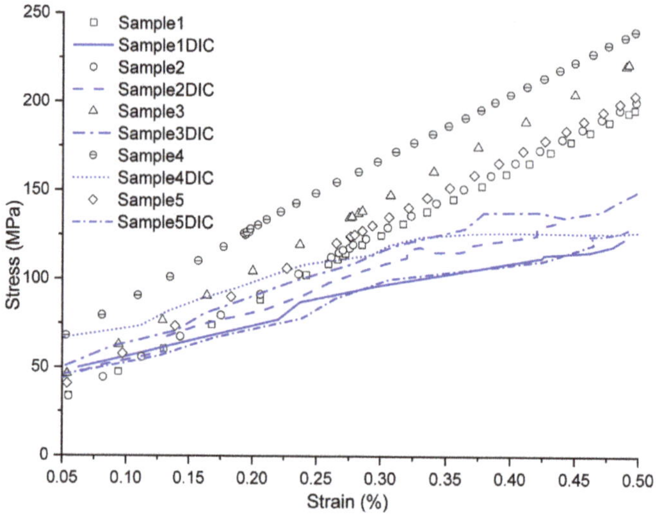

Figure 5. Stress vs. strain graph of the $[0]_3$ configuration, within the range where the analyses were conducted.

Table 2. Average of the properties obtained in the tests compared to the literature.

Mechanical Properties	DIC	Analog Extensometer	Mech-G [30]	Horlle [31]	Fiorelli [32]	Reis [33]
E_1 [MPa]	[1] 34,459 ± 515 (1.5%)	[1] 38,538 ± 1191 (3.1%)	44,816	20,600	29,187	23,930
E_2 [MPa]	[1] 9581 ± 602 (6.3%)	[1] 10,152 ± 757 (7.5%)	12,411	3800	-	7588
ν_{12}	[1] 0.29 ± 0.02 (6.9%)	-	0.28	0.27	0.30	0.30
ν_{21}	-	[2] 0.08 ± 0.01 (12.5%)	-	-	-	-
G_{12} [MPa]	[1] 2058 ± 24.4 (1.2%)	-	5515	-	-	2270
τ_{12} [MPa]	[1] 28.2 ± 1.0 (3.5%)		68.0	-	-	31.4
σ_1^T [MPa]	[1] 567.4 ± 59.7 (10.5%)		1035	288.8	410	445.4
σ_2^T [MPa]	[1] 7.5 ± 0.2 (2.7%)		48	3.9	-	5.2

[1] Mean ± standard deviation (coefficient of variation). [2] Poisson's ratio 21 calculated from the reciprocity equation.

Figure 6a shows all the test specimens from the $[0]_3$ configuration that were subjected to testing. Among them, some specimens were discarded due to slippage in the fixation elements (tabs), which compromised the data analysis. In the remaining specimens, as mentioned earlier, failure occurred consistently, following the longitudinal direction of the fibers.

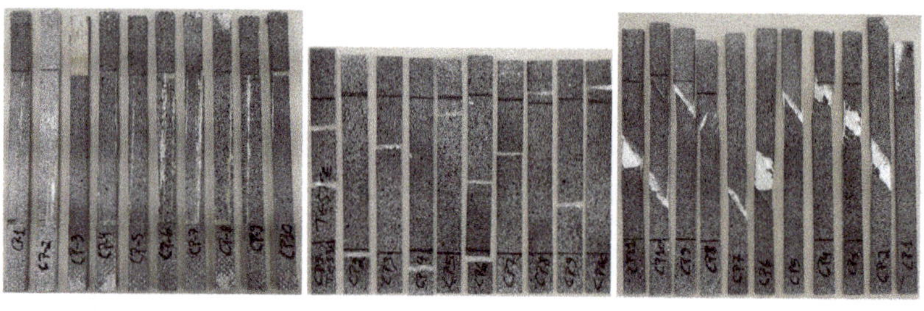

(a) (b) (c)

Figure 6. Samples after mechanical tests: (**a**) $[0]_3$, (**b**) $[90]_4$, and (**c**) $[\pm 45]_3$.

In this configuration, a significant complication was noted in capturing strains due to the alignment of the laminate reinforcement in the same direction as the applied load. This alignment resulted in a stiffer laminate, which made the precise detection of strains challenging.

3.2. Results of Laminate $[90]_4$

The specimens from the $[90]_4$ configuration are depicted in Figure 6b after the tensile tests were conducted. This configuration encountered difficulties regarding the coherence and location of failures in the samples. In some cases, failure occurred within the grip, while, in others, it occurred in the center of the sample.

It was observed that all failures in the specimens occurred in the fiber direction, regardless of where they manifested, considering the displacement speed of the universal testing machine, which was 2 mm/min.

The tests revealed excessively low loads compared to other conditions, due to the transverse alignment of the fibers relative to the applied load. Additionally, the high variability observed in the readings of the analog extensometer was a consequence of the sample's fragility, where small variations in load resulted in significant changes in extensometer readings, making it challenging to obtain consistent and reliable data, particularly due to the high robustness of the machine used relative to the load level.

When comparing this configuration with the previous $[0]_3$, the presence of anisotropy in the composites was noted. Considering it is the same material, the only difference lies in the direction in which the load is applied relative to the fibers, resulting in distinct properties. Moreover, the fact that the $[90]_4$ configuration has an additional reinforcement layer compared to $[0]_3$ suggests that, if both configurations had the same number of reinforcement layers, the discrepancy between the presented values could be even greater.

Regarding the strains obtained by DIC, Figure 4b displays the strain field in the test for a load level of 395 N. It was evident that there was reduced strain in both the longitudinal and transverse extensometers. In this configuration, the camera encountered difficulties capturing strains due to the high sensitivity of the sample, low strains, and limited testing time, resulting in a shorter data recording period compared to other laminates and restricting the detailed analysis of the sample's behavior over time.

In the stress–strain diagram of Figure 7, ten curves representing the laminate are presented, with five captured by the analog extensometer and the other five by the digital extensometer from DIC.

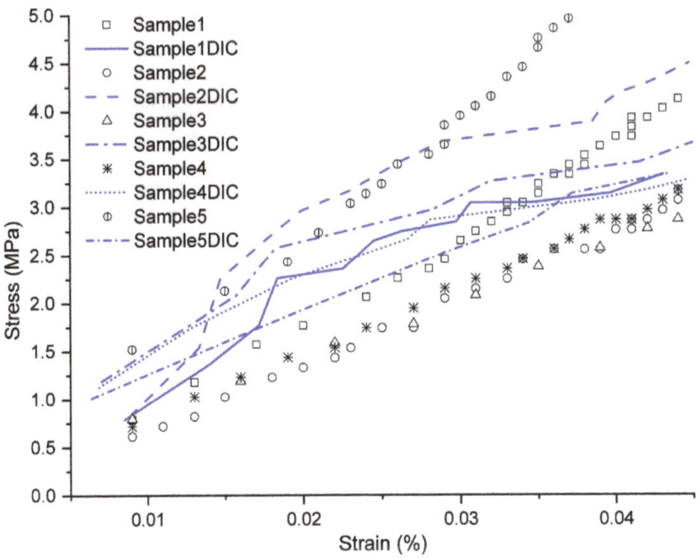

Figure 7. Stress vs. strain graph of the $[90]_4$ configuration, within the range where the analyses were conducted.

Table 2 presented the results of the elastic moduli, one obtained through the analog extensometer and the other with the digital extensometer using the DIC technique. Additionally, the tensile strength values in the transverse direction to the fibers were described, obtained through the load cell of the testing machine, relating the applied force to the area of the sample's cross-section.

In this configuration, a difficulty was encountered due to the high sensitivity of the samples, where the alignment of the fibers transverse to the load amplified this effect, requiring precise detailing within a short testing period.

3.3. Results of Laminate $[\pm 45]_3$

Figure 6c presented the third series of tests on the machine, where all samples failed in a coherent and consistent manner, following the direction of the fibers and accompanied by a reduction in the cross-sectional area, something that was not observed in other configurations. The test naturally demonstrated the highest strain due to the alignment of the fibers at a 45-degree angle to the applied load. Additionally, it was confirmed that these specimens exhibited good shear strength.

During the execution of this test, it is important to highlight that the analog extensometer was used until the deformation reached approximately 1%, as a safety measure to prevent equipment damage in case of unexpected sample failure. The displacement speed of the universal testing machine was 2 mm/min.

For the strains obtained by DIC, as demonstrated in Figure 4c, the strains recorded in both the longitudinal and transverse extensometers were substantially larger compared to previous configurations. This observation makes this laminate particularly interesting for study.

The greater magnitude of strains provides significant advantages in characterizing laminated composite materials. This approach allows for a more robust and detailed data collection, as well as enabling the analysis of the initiation and propagation of progressive damage, which are crucial elements in understanding the behavior of these materials under load. The higher strains recorded by strain gauges provides a broader margin for interpreting the results, facilitating the identification of trends and specific behaviors of the laminate.

The deformation capability of each laminate is inherently related to its stacking sequence, which is essential for distinguishing one laminate configuration from another and for evaluating the material's strength in progressive damage scenarios. The focus of this study lies in analyzing low deformations, despite the laminates being capable of exhibiting higher deformations.

The present study corroborates the findings of Xu et al. [15], as the differences between the deformation obtained by DIC and other methods, such as analytical and numerical, are greater for cases with lower levels of deformation. The analysis of deformations in hybrid composites reinforced with natural fibers, conducted by Xu et al., revealed important insights into the mechanical behavior of these materials under load. Longitudinal and transverse deformations were obtained using DIC for flax/epoxy, jute/epoxy, flax/glass/epoxy, and jute/glass/epoxy composites. The results showed that the addition of glass layers reduced the transverse contraction coefficients compared to composites reinforced only with natural fibers, increasing the difference between the methods. This correlates with the present study, as for the 0 and 90 samples with lower levels of deformation, the difference was greater when compared to the 45 samples.

In Figure 8, ten curves are represented, enabling a comprehensive comparison between the accuracy of DIC and the analog extensometer. Both approaches revealed notable fidelity in capturing strains, given that this configuration had a greater strain than the other conditions, thus facilitating the acquisition of deformation fields. It is observed that the curves obtained with DIC have small markers on their respective curves to reduce visual clutter in the diagram.

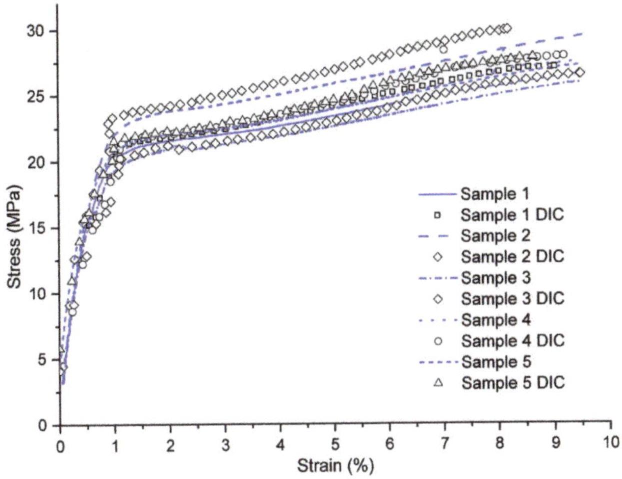

Figure 8. Stress vs. strain graph of the $[\pm 45]_3$ configuration with strain obtained by DIC and analog extensometer.

In Figure 9, using the stress–strain diagram, ten curves are presented to represent the laminate, employing both the analog extensometer method and the digital DIC extensometer method. All curves start with similar stress levels. However, the curves obtained with the analog extensometer remain stable, whereas those obtained with the digital extensometer show slight variations. This is because the analog extensometer recorded data every 0.1 s, whereas the digital method did so every 1 s.

In this specific configuration, the camera used demonstrated a remarkable ability to capture deformations extremely accurately, aligning closely with the data recorded by the analog extensometer. This precision can be mainly attributed to the fact that this specific configuration exhibits higher levels of strain compared to the other conditions. Consequently, the data recording period was relatively longer compared to the other laminates, allowing for a more detailed and precise analysis of the sample's behavior over time.

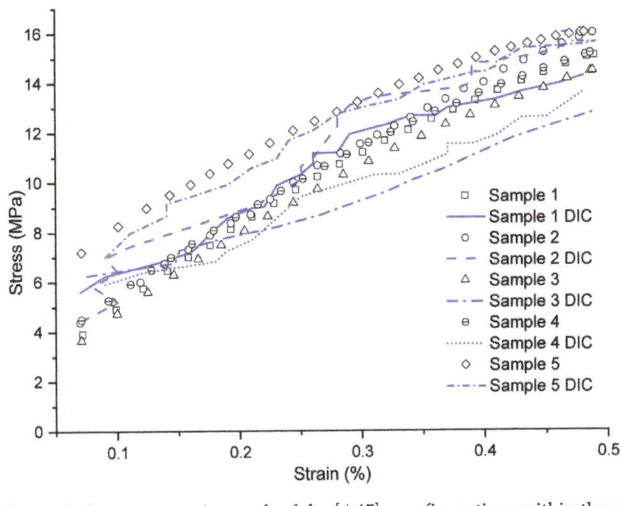

Figure 9. Stress vs. strain graph of the $[\pm 45]_3$ configuration, within the range where the analyses were conducted.

In Table 2, another data point presented is the shear modulus in the 12 plane, which can only be obtained with the assistance of an additional digital extensometer in the transverse direction, requiring the simultaneous use of two extensometers attached to the sample. The values of tensile strength in the longitudinal direction of the laminate are described, as well as the shear strength in the 12 plane, which essentially equals half of that value, both obtained through the testing machine, relating the applied force to the cross-sectional area of the sample. Overall, the calculated coefficient of variation was small, indicating low data dispersion, and thus the number of samples was adequate.

3.4. Discussions of Results

The comparison of the results obtained with the literature in the field was conducted using the data compiled in Table 2. Mechanical properties such as the elastic modulus in directions 1 and 2, shear modulus in the 12 plane, and Poisson's ratio in planes 12 and 21, as well as tensile strength in directions 1 and 2 and shear strength in the plane, were collected based on the average values obtained from a rigorous statistical analysis of the tests, both by the DIC method and the analog method. Additionally, comparisons with the literature were provided to validate the results obtained in the tests.

All cited studies used unidirectional E-glass fabric. In study [31], polyester resin was used, which could be disregarded for comparison with the elastic modulus in the fiber direction, as the reinforcement configuration is the most critical aspect. On the other hand, the studies [32,33] used epoxy resin similar to that tested experimentally in the present study, but from different manufacturers. Other properties were obtained using Mech-G software [30] developed by the Composite and Nanocomposite Materials Group (GCOMP) at LAPOL/UFRGS.

It is important to note that, in all studies, the tensile tests followed the ASTM D3039 standard, ensuring the comparability of the results. Another fundamental point discussed in the studies was the sample manufacturing method, which was mostly carried out through hand lay-up, except for the study by Reis, which employed the vacuum infusion process.

It is worth emphasizing that the manufacturing process of composite materials significantly influences the magnitude of the properties obtained experimentally. The study [34] highlighted that the manufacturing technique used can significantly affect the mechanical properties of composites, including tensile strength and elastic modulus.

In laminate $[0]_3$, the stress–strain curves obtained by the analog extensometer showed an average modulus of elasticity of 38,538 MPa. Conversely, the curves obtained by the digital extensometer exhibited some variations, resulting in an average modulus of elasticity value of 34,459 MPa. This represented a relative difference of 11.1%; in the literature, values obtained in similar studies also varied for this same property.

The Poisson's ratio 12 found was 0.29, very close to the value of 0.28 obtained in the [30] reference. The greatest discrepancy between the results of this test and those from the literature was in the tensile strength in the fiber direction, which was 1035 MPa, contrasting with the 567.4 MPa obtained in the mechanical tests. In the cited experimental studies, the tensile strength limits were 288.8 MPa, 410 MPa, and 445.4 MPa, for [31–33], respectively. These values were more consistent with the magnitude obtained in the tests, suggesting that the number of plies and the manual lamination process interfere with the results obtained.

In laminate $[90]_4$, DIC showed difficulties in capturing strains due to the high sensitivity of the samples. However, it is noteworthy that the obtained data significantly approached the values found in the literature. The results were, respectively, 9581 MPa using DIC, 10,152 MPa with the analog method, and 12,411 MPa in the literature, using [30] as a reference.

Regarding tensile strength in the direction transverse to the fibers, the values were 7.5 MPa obtained in the tests and 48 MPa in the literature. However, comparing the strength obtained in the tests with the epoxy resin strength indicated by the manufacturer [35], which is 7.5 MPa, coherence becomes evident. This is because, in this test, the properties most significantly requested are those of the resin, and the fibers hardly influence the strength in the transverse direction.

In laminate $[\pm45]_3$, comparing the shear modulus properties obtained with the mechanical tests, they were respectively 2058 MPa and 5515 MPa obtained with the Mech-G software, while, in study [33], a value close to 2270 MPa was obtained. Regarding shear strength in plane 12 of the tests of this study, 28.2 MPa was obtained, while, in the literature, it was 68 MPa and 31.4 MPa, for [30,33], respectively.

The values of the data obtained in the tests differed from the indicated studies, mainly due to the number of reinforcement layers used in the composite. Another relevant factor was the type and characteristics of the manufacturing process, including the curing methods of the laminate. Since the data were experimentally obtained, there are unquantified uncertainties that influence the final result. This disparity was understandable, as the materials, despite being the same (unidirectional E-glass fiber and epoxy resin), are from different manufacturers and undergo different processes and treatments, which consequently result in different characteristics and properties obtained for the laminate.

Moreover, the analytical calculation based on classical laminate theory was used to obtain the engineering constants of laminates, including the longitudinal elastic modulus E_x. The data used to feed the method were primarily extracted from Table 2, pertaining to laminates $[0]_3$, $[90]_4$, and $[\pm45]_3$, the latter considering data obtained from local coordinates. Additionally, information such as stacking sequence, average laminate thickness, and number of layers were also incorporated.

Table 3 presents the average values of the results obtained for the engineering constant of the $[\pm45]_3$ laminate, using both the DIC and analog methods, considering the global coordinates of the system.

Table 3. Data obtained in the trials compared to the script.

E_x			Relative Difference [%]	Relative Difference [%]
Analog Extensometer	DIC	Classical Laminated Plate Theory	Theory vs. Analog	Theory vs. DIC
[1] 5208 ± 172 (3.3%)	[1] 4945 ± 243 (4.9%)	7136	31.2	36.2

[1] Mean ± standard deviation (coefficient of variation).

Regarding the modulus of elasticity obtained for this laminate, the results were consistent as analyzed through the classical laminate theory. It is important to note that the analysis was conducted without considering interactions related to progressive damages occurring during the failure of composite materials. In other words, the approach adopted was idealized, resulting in a modulus of elasticity value of 7136 MPa. Conversely, the experimentally obtained value for the modulus of elasticity was 5208 MPa. This discrepancy represents a relative difference of 31.2%, primarily attributed to the manufacturing process involving unquantified uncertainties combined with progressive damage during material failure in the experimental tests.

4. Conclusions

This study investigated the mechanical properties of three composites with different stacking sequences reinforced with E-glass fibers, using DIC measurements and an analog extensometer during tensile tests. The DIC method, performed with simplified equipment, effectively determined the elastic and shear moduli as well as the Poisson's ratio of the analyzed laminates. Additionally, the strain fields obtained by the DIC technique contribute to a better understanding of sensitivity and limitation of the simplified equipment.

Visual analysis of the specimens after testing revealed that failures predominantly occurred at the interface between the fiber and the composite matrix, without rupture of the glass fiber filaments. This failure mechanism was consistent across all examined cases. In the $[0]_3$ laminate, the samples exhibited high mechanical strength, averaging 567.4 MPa, and low strain, with the limited resolution of the equipment detecting minimal strains. In the $[90]_4$ laminate, lower strains and tensile strength were observed, averaging 7.5 MPa, indicating high sensitivity to test conditions. The DIC technique proved effective even under these critical conditions. In the $[\pm45]_3$ laminate, greater strain was noted compared to the other configurations, and the DIC method demonstrated high precision in detecting strains, generating curves similar to those obtained by the analog extensometer.

The comparison between the theoretical and experimental results for the elastic modulus E_x of the $[\pm45]_3$ laminate revealed a 31.2% difference, indicating uncertainties in manufacturing and the absence of considerations about progressive damage in the theoretical analysis. The DIC technique with simplified equipment showed limits in detecting strain in more rigid conditions, with fibers aligned in the same direction as the load. However, for laminates with varied stacking sequences, a simplified DIC can be utilized. This study provides an important reference and basis for future research on the mechanical behavior of fiber-reinforced composites.

Author Contributions: Conceptualization was led by D.Y. and M.L.P.T.; methodology was developed by D.Y.; validation was conducted by D.Y., L.R.C.D., M.L.P.T., A.K. and J.Š.-J.; formal analysis was carried out by D.Y., S.M. and M.L.P.T.; investigation was led by D.Y.; resources were provided by D.Y. and M.L.P.T.; data curation was performed by M.L.P.T.; original draft preparation was carried out by D.Y.; review and editing were conducted by D.Y., S.M., A.K., J.Š.-J. and M.L.P.T.; visualization was performed by M.L.P.T.; supervision was the responsibility of L.R.C.D. and M.L.P.T.; project administration was carried out by M.L.P.T.; and funding acquisition was performed by D.Y., A.K. and J.Š.-J. All authors have read and agreed to the published version of the manuscript.

Funding: This research received no external funding.

Data Availability Statement: The original contributions presented in the study are included in the article, further inquiries can be directed to the corresponding author.

Conflicts of Interest: The authors declare no conflicts of interest.

Abbreviations

The following abbreviations are used in this manuscript:

DIC	Digital Image Correlation
CNC	Computer numerical control
LED	Light-emitting diode
FPS	Frames per second
CFRP	Carbon fiber reinforced polymer

References

1. Cândido, G.M.; Almeida, S.F.M. Processing of Advanced Polymer Composite Laminates with Molded Edges. *Adv. Polym. Compos. Laminates* **2010**, *10*, 31–41.
2. Neto, F.L.; Pardini, L.C. *Structural Composites;* Science and Technology; Edgard Blucher: Sao Paulo, Brazil, 2006.
3. Rezende, M.C. Fractography of Structural Composites. *Polím. Ciênc. Tecnol.* **2007**, *17*, 1–8.
4. Baggio, A.; Moraes Cruz, M.P.; Bento Gonçalves, G.d.M.; Sanchez, L.E.d.Â. Obtaining Low Density Composites and Their Characterization Using the Hand Lay-Up Manual Lamination Process with Unsaturated Polyester Matrix Reinforced with Fiberglass. *Rev. Mater.* **2022**, *27*, 1–14. [CrossRef]
5. Callister, W.D.J.; Rethwisch, D.G. *Materials Science and Engineering: An Introduction*, 9th ed.; LTC: Rio de Janeiro, Brazil, 2016.
6. Jones, R.M. *Mechanics of Composite Materials*, 2nd ed.; Taylor & Francis: London, UK, 1999; Volume 1, ISBN 1715830490.
7. Belaid, S.; Bensmain, B.; Balland, P. Study of Changes of Mechanical Properties of an Aged Composite by Digital Image Correlation. *Recl. Méc.* **2021**, *5*, 503–512. [CrossRef]
8. Kaw, A.K. The title of the cited contribution. In *Mechanics of Composite Materials*, 2nd ed.; CRC Press: New York, NY, USA, 2006; 466p.
9. Neto, J.L.B. Study of the Mechanical Behavior of Polymer Matrix Composites with Fiberglass Containing Geometric Discontinuities. Master's Thesis, Federal University of Campina Grande, Campina Grande, Brazil, 2015.
10. Sabik, A.; Rucka, M.; Andrzejewska, A.; Wojtczak, E. Tensile Failure Study of 3D Printed PLA Using DIC Technique and FEM Analysis. *Mech. Mater.* **2022**, *175*, 80–233. [CrossRef]

11. Costello, T.; Chan, M. Carbon Fiber, One of Titan Submersible's Experimental Materials, Comes under Scrutiny. Available online: https://www.nbcnews.com/news/us-news/carbon-fiber-one-titan-submersibles-experimental-materials-comes-scrut-rcna90856 (accessed on 23 June 2023).
12. Blikharskyy, Y.; Kopiika, N.; Khmil, R.; Selejdak, J.; Blikharskyy, Z. Review of Development and Application of Digital Image Correlation Method for Study of Stress–Strain State of RC Structures. *Appl. Sci.* **2022**, *12*, 157. [CrossRef]
13. de Assis, L.S.; Pessoa, J.R.d.C.; Tavares Júnior, A.D. Elaboration of Fracture Prediction Map Using 2D Digital Image Correlation—2D CID. *Rev. IBRACON Estruturas Mater.* **2022**, *15*, 1–9. [CrossRef]
14. Blenkinsopp, R.; Roberts, J.; Harland, A.; Sherratt, P.; Smith, P.; Lucas, T. A Method for Calibrating a Digital Image Correlation System for Full-Field Strain Measurements during Large Deformations. *Appl. Sci.* **2019**, *9*, 2828. [CrossRef]
15. Xu, D.; Cerbu, C.; Wang, H.; Rosca, I.C. Analysis of the Hybrid Composite Materials Reinforced with Natural Fibers Considering Digital Image Correlation (DIC) Measurements. *Mech. Mater.* **2019**, *135*, 1–11. [CrossRef]
16. Picoy, Y.S.M. Digital Image Correlation for Displacement Measurement in Cantilever Beams. Master's Thesis, Federal University of Lavras, Lavras, Brazil, 2016.
17. Gonabadi, H.; Oila, A.; Yadav, A.; Bull, S. Investigation of Anisotropy Effects in Glass Fibre Reinforced Polymer Composites on Tensile and Shear Properties Using Full Field Strain Measurement and Finite Element Multi-Scale Techniques. *J. Compos. Mater.* **2021**, *56*, 507–524. [CrossRef]
18. Feito, N.; Calvo, J.V.; Belda, R.; Giner, E. An Experimental and Numerical Investigation to Characterize an Aerospace Composite Material with Open-Hole Using Non-Destructive Techniques. *Sensors* **2020**, *20*, 4148. [CrossRef] [PubMed]
19. Hao, W.; Liu, Y.; Wang, T.; Guo, G.; Chen, H.; Fang, D. Failure Analysis of 3D Printed Glass Fiber/PA12 Composite Lattice Structures Using DIC. *Compos. Struct.* **2019**, *225*, 1–12. [CrossRef]
20. Godara, A.; Raabe, D. Influence of fiber orientation on global mechanical behavior and mesoscale strain localization in a short glass-fiber-reinforced epoxy polymer composite during tensile deformation investigated using digital image correlation. *Compos. Sci. Technol.* **2007**, *67*, 2417–2427. [CrossRef]
21. Namala, K.K.; Mahajan, P.; Bhatnagar, N. Digital Image Correlation of Low-Velocity Impact on a Glass/Epoxy Composite. *Int. J. Comput. Methods Eng. Sci. Mech.* **2014**, *15*, 203–217. [CrossRef]
22. Maneghi, S.; Sarwar, A.; Fawaz, Z.; Zdero, R.; Bougherara, H. Mechanical characterization of the static and fatigue compressive properties of a new glass/flax/epoxy composite material using digital image correlation, thermographic stress analysis, and conventional mechanical testing. *Mater. Sci. Eng. C* **2019**, *99*, 940–950. [CrossRef] [PubMed]
23. Jefferson, A.J.; Arumugam, V.; Santulli, C.; Jennifers, A.; Poorani, M. Failure Modes of GFRP After Multiple Impacts Determined by Acoustic Emission and Digital Image Correlation. *J. Eng. Technol.* **2015**, *6*, 29–35.
24. Salian, A.; Ashwini, T.P. Comparative Study of Strain Using 2D Digital Image Correlation and Extensometer on Glass Fiber. In *Smart Sensors Measurements and Instrumentation*; Santhosh, K.V., Rao, K.G., Eds.; Springer: Singapore, 2021; pp. 351–366. [CrossRef]
25. Bárnik, F.; Sága, M.; Vaško, M.; Handrik, M.; Kopas, P.; Štalmach, O.; Dorčiak, F. Measurement and Comparative Study of Deformations Using Strain Gauge and 2D DIC Technology. *IOP Conf. Ser. Mater. Sci. Eng.* **2020**, *776*, 012065. [CrossRef]
26. Texxiglass. Properties of Fiberglass. Available online: http://texiglass.com.br/wp-content/uploads/2022/05/Propriedades-da-Fibra-de-vidro.pdf (accessed on 27 April 2023).
27. *ASTM D3039/D3039M-17*; Standard Test Method for Tensile Properties of Polymer Matrix Composite Materials. ASTM: West Conshohocken, PA, USA, 2014; pp. 1–13. Available online: https://www.astm.org/d3039_d3039m-08.html (accessed on 9 June 2024).
28. *ASTM D3518/D3518M-18*; Standard Test Method for In-Plane Shear Response of Polymer Matrix Composite Materials by Tensile Test of a ±45° Laminate; ASTM: West Conshohocken, PA, USA, 2018; pp. 1–7. Available online: https://www.astm.org/d3518_d3518m-18.html (accessed on 9 June 2024).
29. GOM GmbH. GOM Correlate: Digital Image Correlation and Evaluation Software. Available online: https://www.gom.com/3d-software/gom-correlate.html (accessed on 25 June 2023).
30. Composite Materials and Nanocomposites Group (GComp)—URGS. Properties of Fiberglass-E Composite and Epoxy Resin. Available online: https://gcomp-srv01.nuvem.ufrgs.br/pt-br/macro/laminae/laminae (accessed on 16 May 2023).
31. Hörlle, F.O. Comparative Evaluation of Unidirectional Polyester Composites with Glass or Curauá Fibers. Master's Thesis, Federal University of Rio Grande do Sul, Porto Alegre, Brazil, 2015.
32. Fiorelli, J.; Dias, A.A. Determination of Mechanical Properties of Strength and Stiffness of Unidirectional Fabrics of Glass and Carbon Fibers. In Proceedings of the Brazilian Congress of Engineering and Materials Science 2012, Joinville, Brazil, 4–8 November 2012; pp. 1651–1657.
33. Reis, L.R. Obtaining and Characterization of Unsaturated Polyester Resin Composites Molded by the Infusion Process. Master's Thesis, Federal University of Itajubá, Itajubá, Brazil, 2016.

34. El Sawi, I.; Bougherara, H.; Zitoune, R.; Fawaz, Z. Influence of the Manufacturing Process on the Mechanical Properties of Flax/Epoxy Composites. *J. Biobased Mater. Bioenergy* **2014**, *8*, 69–76. [CrossRef]
35. E-Composites. AR320LV and AH320 Epoxy Resin System. Available online: https://www.marinecomposites.com.br/wp-content/uploads/datasheets/resina_epoxy_ar320lv.pdf (accessed on 27 April 2023).

materials

MDPI

Article

Compressive Properties and Energy Absorption Characteristics of Co-Continuous Interlocking PDMS/PLA Lattice Composites

Han Wang [1,*], Kedi Wang [1], Jincheng Lei [2] and Xueling Fan [1,*]

1 Xi'an Key Laboratory of Extreme Environment and Protection Technology, School of Aerospace Engineering, Xi'an Jiaotong University, Xi'an 710049, China

2 International Center for Applied Mechanics, State Key Laboratory for Strength and Vibration of Mechanical Structures, Xi'an Jiaotong University, Xi'an 710049, China

* Correspondence: whishappy@foxmail.com (H.W.); fanxueling@mail.xjtu.edu.cn (X.F.)

Abstract: Co-continuous interlocking lattice structures usually present superior compressive properties and energy absorption characteristics. In this study, co-continuous interlocking polydimethylsiloxane/polylactic acid (PDMS/PLA) lattice composites were designed with different strut diameters, and successfully manufactured by combining the fused deposition modeling (FDM) technique and the infiltration method. This fabrication method can realize the change and control of structure parameters. The effects of the strut diameter on the compressive properties and energy absorption behavior of PDMS/PLA lattice composites were investigated by using quasi-static compression tests. The compressive properties of the co-continuous interlocking PDMS/PLA lattice composites can be adjusted in a narrow density range by a linear correlation. The energy absorption density of the co-continuous interlocking PDMS/PLA lattice composites increases with the increase in the PLA strut diameter and presents a higher efficiency peak and wider plateau region. The PLA lattice acts as a skeleton and plays an important role in bearing the compressive load and in energy absorption. The indexes of the compressive properties/energy absorption characteristics and PLA volume fraction of co-continuous interlocking PDMS/PLA lattice composites show linear relationships in logarithmic coordinates. The effect of the PLA volume fraction increasing on the plateau stress is more sensitive than the compressive strength and energy absorption density.

Keywords: compressive properties; energy absorption characteristics; lattice structure; co-continuous composites; interlocking structure

1. Introduction

The materials and structures observed in nature often showcase many captivating architectural designs and exceptional characteristics [1–3]. Classical porous materials, including honeycomb structures and foam structures, show unique properties such as low density and high specific strength and stiffness [4–6]. Inspired by nature, porous structures find applications in vehicles, aircrafts, biological scaffolds and other fields [7,8]. Besides superior mechanical properties, special functional properties such as energy absorption, sound absorption, electromagnetic shielding, damping, permeability and biocompatibility render porous materials highly promising for practical applications [9–13].

Currently, the investigations of lattice structures have received extensive attention [14–17]. In comparison to foam structures with stochastic cells, lattice structures offer the advantages of optimal design and precise control, enabling them to meet the requirements for the special and adjustable properties [18]. This is precisely why lattice structures have garnered significant attention from researchers. Additionally, the rapid advancement of additive manufacturing has led to the widespread use of lattice structures as one of the most commonly employed porous architectures. Nonetheless, the design of a lattice structure alone cannot meet the higher requirements of energy absorption targets like vehicle collision and armor protection.

There is a view that introducing the second solid phase, instead of the gas phase of a lattice structure, is believed to offer a combination of the advantages of the filler material and the original lattice skeleton, leading to a significant improvement in mechanical properties [19–22]. Hence, the co-continuous composites achieved through this method [23–26] are proposed, and the enhancement objective of mechanical properties is successfully achieved.

The co-continuous composites typically consist of two distinct materials in order to allow for the adjustment of the global properties of co-continuous composites. The filler materials must satisfy specific conditions in terms of preparation method feasibility and enhancement of mechanical properties when compared to the original lattice skeleton [21]. Therefore, the two phases of co-continuous composites are usually chosen from metals, ceramics, polymers and even liquid phase materials. Wang et al. [23] demonstrated the potential to design and fabricate co-continuous glassy polymer/rubbery polymer materials with enhancements in stiffness, strength and energy dissipation using a 3D printer. Three types of co-continuous structures with simple cubic (SC), body-centered cubic (BCC) and face-centered cubic (FCC) Bravais lattices were considered in their study. Liu et al. [24] investigated the quasi-static compressive behaviors of the co-continuous glassy polymer/liquid composites. Due to the presence of liquid filler, the stiffness, yield strength and energy absorption of co-continuous composites were significantly enhanced, which comes from the additional support of the liquid filler and the lateral expansion of the glassy polymer. Geometric structures of the glassy polymer, namely a simple cubic (SC) lattice, a face-centered cubic (FCC) lattice and a body-centered-cubic (BCC) lattice, were investigated. Al-Ketan et al. [27,28] employed 3D printing technology to fabricate interpenetrating phase composites (IPCs) and investigated the mechanical properties of IPCs with different periodic architectures comprising a soft matrix and a continuous, smooth-curved, hard material. Additionally, Al-Ketan et al. [25] proposed a nature-inspired 3D periodic shell–core cellular co-continuous composites made of a hard-shell and soft-core system based on a Gyroidal triply periodic minimal surface. Mansouri et al. [29] investigated the mechanical behavior of a 3D periodic single-material cellular D-structure and the corresponding co-continuous composite combining hard thermoplastic materials with soft rubbers with a rupture strain of more than 500%, fabricated using multimaterial fused deposition modeling (FDM) technology. Wang et al. [21] prepared $Mg_{17}Al_{12}/Al$ ordered structure composites by infiltrating the ordered porous aluminum with the intermetallic compound $Mg_{17}Al_{12}$ under gravity conditions. The intermetallic compound $Mg_{17}Al_{12}$ exhibits significant brittleness while the ordered porous aluminum shows excellent plasticity and ductility. Chen et al. [30] propose to introduce two-phase co-continuous composites as acoustic metamaterials. The two phases of co-continuous composites were chosen to be a ceramic (boron carbide) and a glassy polymer (epoxy). The aforementioned analysis indicates that co-continuous composites have the potential to enhance mechanical properties and optimize functional properties, regardless of the chosen preparation method. The changes in the deformation mechanism and failure mode caused by the presence of the co-continuous interpenetrating phase should be taken into consideration, and the effect of co-continuous interpenetrating structure parameters on mechanical properties should be investigated.

In this study, polylactic acid (PLA) lattices with simple cubic structure were fabricated using the FDM technique as an infiltration skeleton and then filled with liquid polydimethylsiloxane (PDMS) under gravity conditions. The co-continuous interlocking PDMS/PLA lattice composites were prepared following a combination of the FDM technique and the infiltration method. The effect of the strut diameter on the compressive properties and energy absorption behavior of the PDMS/PLA lattice composites were investigated using quasi-static compression tests. The cooperative deformation mechanism and failure mode of the co-continuous interlocking PDMS filler and PLA matrix skeleton were analyzed. The objective of this study was to demonstrate that modifying the parameters of the co-continuous interlocking structure could effectively enhance compressive properties and energy absorption characteristics.

2. Materials and Methods

2.1. Design of Co-Continuous Interlocking PDMS/PLA Lattice Composites

The co-continuous interlocking lattice structures in this study were designed and modeled using the software SOLIDWORKS 2018. The PLA lattice structures consisted of several orthogonal cylinders with different strut diameters based on previous studies [16]. Figure 1 shows the 2D/3D PLA lattice structures, the 3D co-continuous interlocking PDMS/PLA lattice composites and their corresponding structure parameters including the strut diameter a (3, 5, 7 mm), the strut spacing d (12 mm), the porosity P and the volume fraction V_f. The overall dimensions of the CAD models are 36 mm × 36 mm × 36 mm. The volume fractions V_f of the PLA lattice A, B and C are 12.5%, 30.7% and 52.1%, respectively. The corresponding porosities are 87.5%, 69.3% and 47.9%, namely the volume fractions of the PDMS. With the strut diameter increasing, the volume fractions V_f of the PLA increase and the volume fractions of the PDMS decrease.

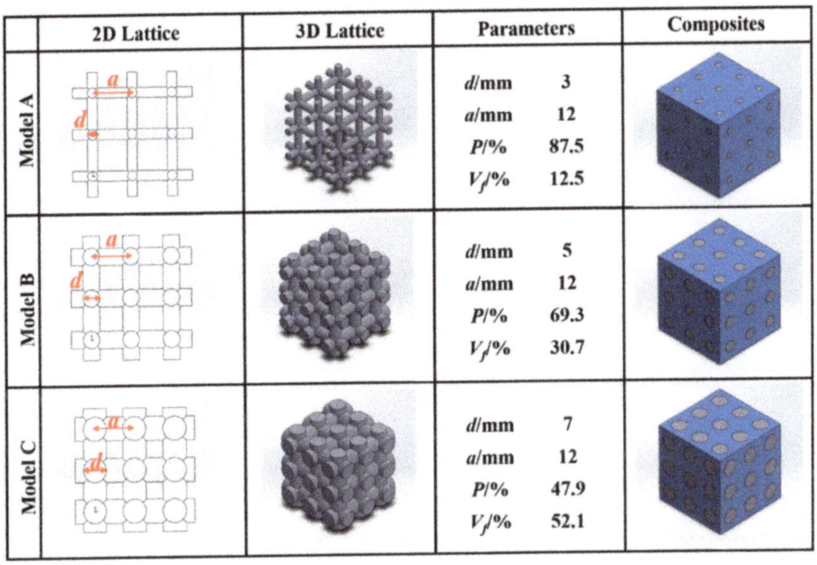

Figure 1. Illustration of 2D/3D PLA lattice structures, corresponding structure parameters and 3D co-continuous interlocking PDMS/PLA lattice composites.

2.2. Preparation of Co-Continuous Interlocking PDMS/PLA Lattice Composites

The PLA lattice structures were prepared using a 3D printer (Creator-pro, Hangzhou, China) with the FDM technique [31,32] in this investigation. The raw material of the 3D printer is a commercial wire PLA product (FLASHFORGE, Hangzhou, China). The models of the PLA lattice structures were sliced and imported into the 3D printer. The corresponding 3D printing parameters are as follows: slice thickness, 0.12 mm, nozzle preheating temperature, 200 °C, printing speed, 50 mm/s, printing accuracy, ±0.4 mm, etc. After the 3D printing process, the extra support materials were removed and the PLA lattice structures were cleaned manually.

Figure 2 shows the fabrication process of the co-continuous interlocking PDMS/PLA lattice composites. The 3D-printed PLA lattice structure was set as an infiltration skeleton and placed in the corresponding infiltration mold (FLASHFORGE, Hangzhou, China). The PLA lattice structure was in contact with the four internal surfaces of the mold and surrounded by the mold. The PDMS (Sylgard 184, Tianjin, China) was prepared by mixing component A (monomers, Sylgard 184, Tianjin, China) and component B (crosslinkers, Sylgard 184, Tianjin, China) with the mass ratio of A:B = 10:1. Component A was poured into a beaker and then component B was dripped into component A. The mixture was stirred for 15 minutes and vacuumed to remove air bubbles from the solution. The PDMS was then utilized to infiltrate the PLA lattice structure in the infiltration mold under a slight vibration at room temperature. After the PDMS infiltration process, the PDMS/PLA mixture was placed into a vacuum furnace (SHKTYQ, Shanghai, China) and heated at 50 °C for 4 hours. At this temperature, the PDMS cross-linking and curing reaction occurred. Eventually, the external mold was removed and the specimens of co-continuous interlocking PDMS/PLA lattice composites were successfully prepared. The fabrication process of the co-continuous interlocking PDMS/PLA lattice composites consists of two main stages, 3D printing and infiltration casting. As for limitations in 3D printing, there may be printing size errors or anisotropy. As for limitation in infiltration casting, the infiltration of filler may introduce infiltration defects, such as a small number of bubbles.

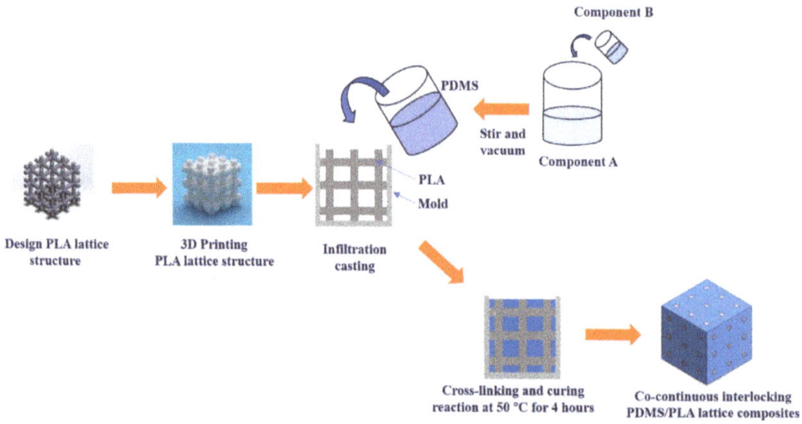

Figure 2. The fabrication process of co-continuous interlocking PDMS/PLA lattice composites.

2.3. *Characterization of Co-Continuous Interlocking PDMS/PLA Lattice Composites*

All specimens of the PLA lattice structures and co-continuous interlocking PDMS/PLA lattice composites were photographed with a digital camera (Nikon, S6300, Tokyo, Japan) and weighed with an analytical electronic balance. The densities of specimens were calculated according to the mass and structure parameters. The quasi-static compression tests were performed on an electromechanical universal mechanical testing machine (MTS, C45.105, Eden Prairie, MN, USA) at an initial strain rate of 10^{-3} s^{-1} to evaluate the compressive properties and energy absorption behavior of the co-continuous interlocking PDMS/PLA lattice composites. To ensure data stability and repeatability, at least three compressive specimens were tested for each condition at room temperature. The compressive strength, plateau stress, densification strain, energy absorption density and energy absorption efficiency were calculated. Macro images of specimens at different compressive strains were recorded during the quasi-static compression tests. The deformation and failure processes of the co-continuous interlocking PDMS/PLA lattice composites were analyzed.

3. Results and Discussion

3.1. *Structure Characterization of Co-Continuous Interlocking PDMS/PLA Lattice Composites*

The specimens of the PLA lattice structures and co-continuous interlocking PDMS/PLA lattice composites are shown in Figure 3. The PLA lattice exhibits a basic truss structure with different strut diameters of 3, 5 and 7 mm. It is observed that the specimens of the co-continuous interlocking PDMS/PLA lattice composites consist of two parts, namely the PLA lattice structure as a skeleton and the PDMS as a filler. These two components become an integral structure with co-continuous interpenetrating phases. The co-continuous interpenetrating structure characteristic of the PDMS and PLA can be visually seen from the specimens' appearance due to the transparent features of the PDMS. The changes in strut diameter and the infiltration treatment cause variations of specimen mass and density, which should be considered when evaluating mechanical properties and energy absorption characteristics. The specimen label, filler, strut diameter, mass and density are shown in Table 1. Compared with the specimens made by Al-Ketan et al [25,27,28], Wang et al. [26] and Mansouri et al. [29], the specimens in this study avoided the filler anisotropy caused by 3D printing. Compared with the specimens made by Wang et al. [21], the specimens in this study show fewer defects due to the longer infiltration time.

Figure 3. The specimens of PLA lattice structures and co-continuous interlocking PDMS/PLA lattice composites.

Table 1. The specimen label, filler, strut diameter, mass and density of the PLA lattice structure and co-continuous interlocking PDMS/PLA lattice composites.

Specimen Label		Filler	d/mm	m/g	ρ/g·cm^{-3}
1	PLA-3	/	3	7.151 ± 0.018	0.153 ± 0.000
2	PDMS/PLA-3	PDMS	3	49.743 ± 0.142	1.066 ± 0.003
3	PLA-5	/	5	17.358 ± 0.157	0.372 ± 0.003
4	PDMS/PLA-5	PDMS	5	50.902 ± 0.176	1.091 ± 0.004
5	PLA-7	/	7	29.750 ± 0.074	0.638 ± 0.002
6	PDMS/PLA-7	PDMS	7	53.346 ± 0.285	1.143 ± 0.006

3.2. Compressive Properties of Co-Continuous Interlocking PDMS/PLA Lattice Composites

The compressive stress–strain curves of co-continuous interlocking PDMS/PLA lattice composites PDMS/PLA-3, PDMS/PLA-5 and PDMS/PLA-7 are shown in Figure 4. For each type of specimen, three compressive stress–strain curves are plotted here. For the PDMS/PLA-3 specimens, a prominent linear elastic stage appears initially followed by an obvious stress peak. The stress declines quickly after reaching its peak, indicating the occurrence of a stress drop. Subsequently, the stress enters a second stage, namely the plateau region. It is found that the stress in this region fluctuates. Finally, the stress increases rapidly again and enters the densification stage. The compressive stress–strain curves of the PDMS/PLA-5 specimens show different characteristics. First, the stress levels of the PDMS/PLA-5 specimens are obviously higher than those of the PDMS/PLA-3 specimens because of the larger effective cross-section area to bear load. Second, there is no stress drop after the linear elastic stage. Instead, the stress enters a plateau region directly. This indicates that the PDMS/PLA-3 specimens show buckling deformation behavior because of the relatively slenderer strut, while the PDMS/PLA-5 specimens show compressive deformation behavior. Therefore, the plateau stress of the PDMS/PLA-5 specimens is approximately close to the compressive strength. After the plateau region, the stress decreases slowly to the minimum value. Eventually, the stress increases quickly again and the densification stage begins. For the PDMS/PLA-7 specimens, the compressive stress–strain curves still comprise three stages with typical characteristics. The stress level of the whole compression process is further increased compared with the PDMS/PLA-3 specimens and PDMS/PLA-5. For the PDMS/PLA-7 specimens, there is no sudden stress drop like for the PDMS/PLA-3 specimens. The stress is maintained near the plateau stress and then decreases slowly to a stress valley at around strain 0.6. It is worth noting that the width of the plateau region with high and stable stress is wider than that of the PDMS/PLA-3 specimens and PDMS/PLA-5. At the same time, the stress fluctuations are relatively more moderate. The strut diameters of PDMS/PLA-7 are larger and the specimens exhibit a stronger capacity of bearing loads. Figure 4d shows the comparison of stress–strain curves among the co-continuous interlocking PDMS/PLA lattice composites PDMS/PLA-3, PDMS/PLA-5 and PDMS/PLA-7. Each curve was acquired by averaging by the three repeated experiments. It can be directly seen that the stress level is improved significantly. The compressive strength and the plateau stress increase with the strut diameter increasing from 3 to 7 mm.

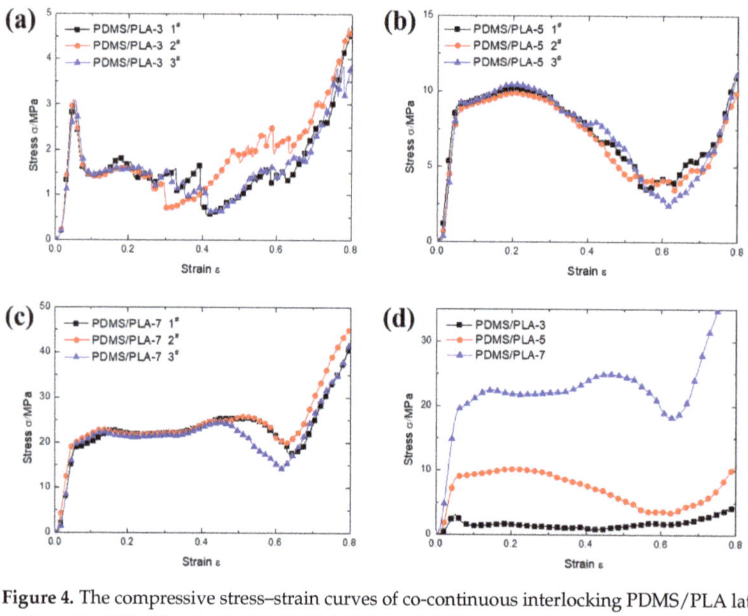

Figure 4. The compressive stress–strain curves of co-continuous interlocking PDMS/PLA lattice composites (**a**) PDMS/PLA-3, (**b**) PDMS/PLA-5, (**c**) PDMS/PLA-7 and (**d**) the comparison among them.

The quasi-static axial compression tests of the co-continuous interlocking PDMS/PLA lattice composites PDMS/PLA-3, PDMS/PLA-5 and PDMS/PLA-7 at different compressive strains (0, 0.1, 0.2, 0.3, 0.4, 0.5, 0.6 and 0.7) are shown in Figure 5. According to Maxwell's criterion, the dominated mechanism of simple cubic structure is the stretching-dominated mechanism. For the PDMS/PLA-3 specimens, the PLA lattice as the skeleton played a main role in bearing the compressive load due to the higher Young's modulus. At the initial stage of compression deformation, the PLA lattice showed linear elastic deformation. With the compressive load increasing, the struts buckled, resulting from the high ratio of length to diameter. At the plateau region, the PLA struts were compressed and bended. Even the joints cracked due to the large stress concentration. Eventually, the PLA struts ruptured and peeled from the specimen. The PLA lattice collapsed layer by layer and entered the densification stage. During the whole compression process, the PDMS acted as a flexible material and wrapped the PLA strut. The compression and lateral expansion of the PDMS filler occurred, hindering the deformation of the PLA. This made a difference in the deformation process of the PLA lattice. The exitance of the PDMS as a filler slows the tendency of the PLA strut to deform and fracture, reflecting that the co-continuous interlocking structure contributes to the improvement of global mechanical properties.

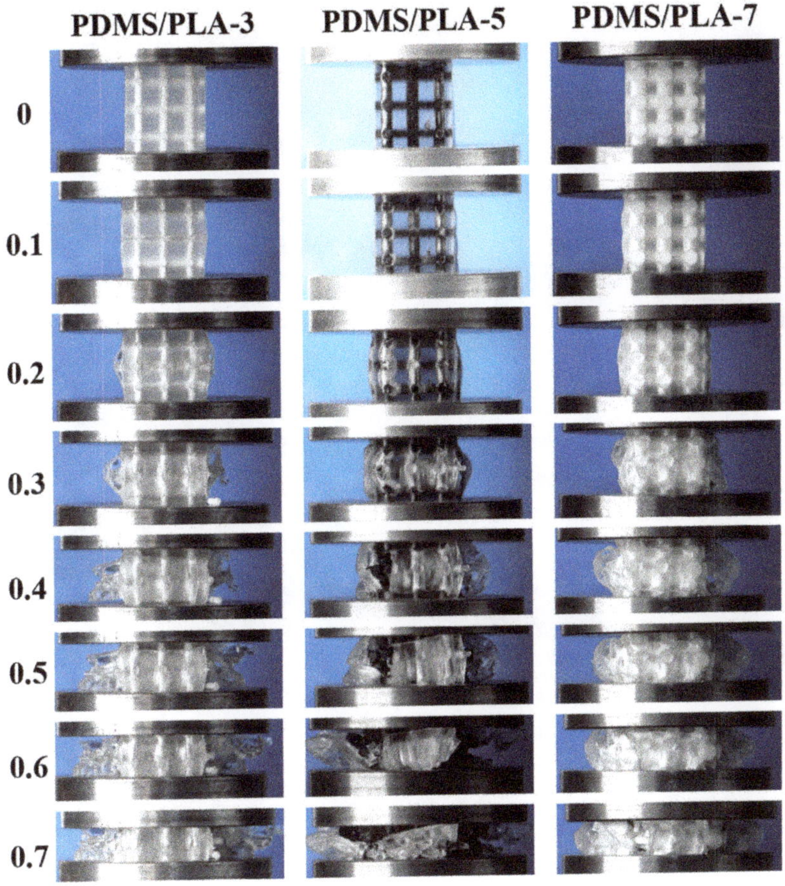

Figure 5. The compressive stress–strain curves of co-continuous interlocking PDMS/PLA lattice composites PDMS/PLA-3, PDMS/PLA-5, and PDMS/PLA-7.

Compared with the compression process of the PDMS/PLA-3 specimen in Figure 5, the compression deformation processes of the PDMS/PLA-5 specimens and PDMS/PLA-7 are similar. Initially, the whole specimen deformed linearly. Subsequently, under the influence of the compression load, the PLA lattice ruptured and the PDMS in the middle of the specimen was deformed by expansion at the plateau region. Finally, the specimen entered the densification stage and more PDMS was expelled from the PLA lattice. It is worth noting that the strut diameters of the PDMS/PLA-5 specimens and PDMS/PLA-7 are larger than that of specimen PDMS/PLA-3. Therefore, the PLA lattice played a more important role in bearing compressive loads. Similarly, the stretching-dominated mechanism was the main deformation mechanism. The struts wrapped by PDMS filler exhibited more superior mechanical properties and stronger stability. This characteristic proves the advantages of a co-continuous interlocking structure. Meanwhile, increasing the strut diameter of the PLA skeleton is beneficial to improve the mechanical properties.

The compressive property indexes of the co-continuous interlocking PDMS/PLA lattice composites are shown in Figure 6. The compressive property indexes include the compressive strength σ_c, the specific compressive strength σ_c/ρ, the plateau stress σ_p and the specific plateau stress σ_p/ρ. The compressive strength is usually taken as the first stress peak of the compressive stress–strain curve. If there no local maximum occurs, the compressive strength is determined at the 1% strain offset. The plateau stress is usually calculated by averaging the stress from strain 0.2 to strain 0.4. For specimen PDMS/PLA-3, the compressive strength was 3.08 MPa and the plateau stress was 1.28 MPa. It was found that the plateau stress was 41.56% of the compressive strength. The lowest plateau stress is not conducive to bearing loads and absorbing energy. The PLA lattice with 3 mm struts as the skeleton of the co-continuous interlocking structure showed poor load-carrying capacity. Although there is more PDMS as filler, the strut diameter of specimen PDMS/PLA-3 is too small. Therefore, the global mechanical properties are inferior. The PDMS/PLA-5 specimen showed higher compressive strength (9.01 MPa) and plateau stress (9.20 MPa). With the increase in strut diameter (5 mm), the compressive strength and the plateau stress increased around 1.93 times and 6.19 times compared with those of the PDMS/PLA-3 specimen, respectively. It was found that the compressive strength and the plateau stress are almost identical. This result is beneficial to bearing load and absorbing energy. The PDMS/PLA-7 specimen exhibited the highest compressive strength (19.55 MPa) and plateau stress (22.20 MPa) in this study. Though the PDMS/PLA-7 specimen has a similar structure to PDMS/PLA-3 and PDMS/PLA-5, it has the largest strut diameter (7 mm) for the PLA lattice. The compressive strength and the plateau stress were 2.17 times and 2.41 times of the PDMS/PLA-5 specimen, respectively. Meanwhile, the plateau stress was close to the compressive strength of specimen PDMS/PLA-5. Considering the wider plateau region, the PDMS/PLA-7 specimen showed a superior capacity of bearing load and absorbing energy.

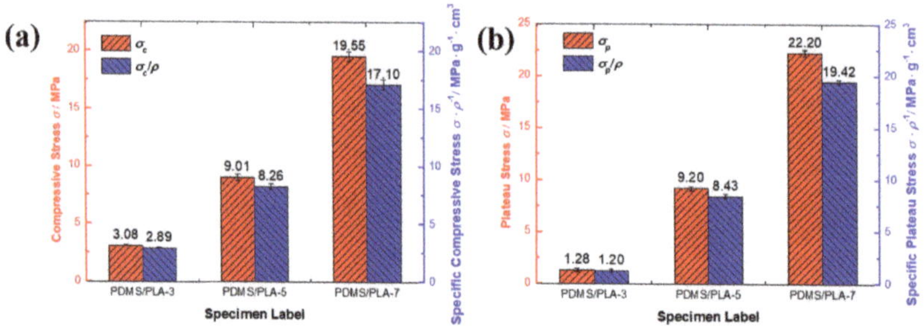

Figure 6. The compressive property indexes: (**a**) compressive strength; (**b**) plateau stress of co-continuous interlocking PDMS/PLA lattice composites.

The increase in strut diameter affects the density of the co-continuous interlocking PDMS/PLA. Considering the effect of the density, the specific compressive strength and the specific plateau stress of the co-continuous interlocking PDMS/PLA were calculated. Due to all co-continuous interlocking PDMS/PLA specimens' density being higher than 1.0 g/cm^3, the specific compressive strength and specific plateau stress slightly decreased compare with the compressive strength and plateau stress. With the strut diameter increasing, the specific compressive strength and specific plateau stress increased. Similarly, the PDMS/PLA-7 specimen still exhibited the highest specific compressive strength and the specific plateau stress in this study. This proves that the strut diameter of the PLA lattice as a skeleton is an important factor that can adjust the mechanical properties of the co-continuous interlocking PDMS/PLA. Figure 7 shows the compressive strength/plateau stress and corresponding density ρ of the co-continuous interlocking PDMS/PLA lattice composites. It was found that the compressive strength and the density are linearly correlated. Similarly, the plateau stress and the density are linearly correlated. The slope (208.16) of the compressive strength and density is slightly lower than that of the plateau stress and density (264.08), proving that the variation in density has a greater effect on the plateau stress of the co-continuous interlocking PDMS/PLA lattice composites.

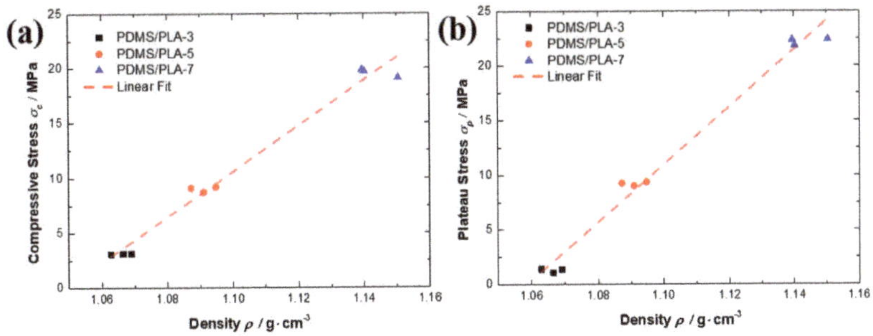

Figure 7. The compressive strength (**a**)/plateau stress (**b**) and corresponding density ρ of the co-continuous interlocking PDMS/PLA lattice composites.

3.3. Energy Absorption Behavior of Co-Continuous Interlocking PDMS/PLA Lattice Composites

The energy absorption density is an important index used to characterize the energy absorption behavior of co-continuous interlocking PDMS/PLA lattice composites, reflecting the energy absorption capacity of materials or structures. The energy absorption density W_V of co-continuous interlocking PDMS/PLA lattice composites can be calculated by Equation (1):

$$W_V = \int_0^{\varepsilon} \sigma(\varepsilon) d\varepsilon \tag{1}$$

where ε and σ are the strain and corresponding stress, respectively. The W_V-ε curves of the co-continuous interlocking PDMS/PLA lattice composites are plotted in Figure 8a. These curves are calculated using the average stress–strain curves for each condition. It is found that the enhancement of energy absorption density is significant with the increase of the strut diameter. For the PDMS/PLA-3 specimen, the W_V-ε curves show a lower level of energy absorption during the whole compression process. The reason is that the PDMS/PLA-3 specimen exhibits a lower compressive stress level, resulting in a lower integral of stress on the strain. When the strut diameter increases to 5 mm, the energy absorption density is obviously improved. During the whole compression process, the energy absorption density of the PDMS/PLA-5 specimen is always higher than that of the PDMS/PLA-3 specimen. The higher stress level and the wider plateau region are both the reason for this result. The PDMS/PLA-7 specimen exhibits the highest level of energy absorption during the whole compression process. Even the energy absorption density of the PDMS/PLA-7 specimen is higher than the sum of the PDMS/PLA-3 specimen and the PDMS/PLA-5 specimen. It is observed that the effect of the strut diameter on energy absorption is significant for the co-continuous interlocking PDMS/PLA lattice composites. The PDMS/PLA-7 specimen has the largest strut diameter, resulting in the widest plateau region with largest plateau stress. These factors lead to the highest energy absorption density in this study.

The energy absorption efficiency reflects the ratio of the energy absorbed by a real material under any strain during compression to the energy absorbed by an ideal energy absorption material under the same strain. The energy absorption efficiency is another important index used to characterize the energy absorption behavior of co-continuous interlocking PDMS/PLA lattice composites. The energy absorption efficiency η of co-continuous interlocking PDMS/PLA lattice composites is calculated by Equation (2):

$$\eta = \frac{\int_0^{\varepsilon} \sigma(\varepsilon) d\varepsilon}{\sigma_{max}\varepsilon} \tag{2}$$

where σ, ε and σ_{max} are the stress, the strain and the maximum compressive stress in the strain range [0, ε], respectively. The η-ε curves of the co-continuous interlocking PDMS/PLA lattice composites are plotted in Figure 8b. These curves are calculated by the average stress–strain curves for each condition.

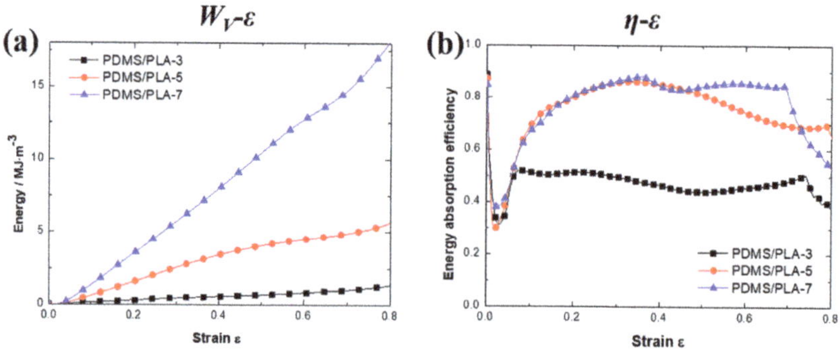

Figure 8. The W_V-ε curves (**a**) and the energy absorption efficiency η-ε curves (**b**) of co-continuous interlocking PDMS/PLA lattice composites.

As for the energy absorption efficiency, these specimens of the co-continuous interlocking PDMS/PLA lattice composites present an obvious difference. For the PDMS/PLA-3 specimen, the energy absorption efficiency rises to about 0.5 at the initial stage. Then, a wide and stable region appears. At this stage, the energy absorption efficiency is maintained at the level of 0.5. Finally, the energy absorption efficiency enters a stage of rapid decline. The η-ε curve of the PDMS/PLA-5 specimen presents a long and gentle rise. Then, it enters a relatively narrow stable region and reaches the maximum. Unfortunately, the maximum efficiency cannot be maintained for a long time with the strain increasing. After the efficiency stable region, the energy absorption efficiency decays slowly to around 0.7. On the whole, the energy absorption efficiency of the PDMS/PLA-5 specimen is higher than that of the PDMS/PLA-3 specimen. It is worth noting that the energy absorption efficiency of specimen PDMS/PLA-7 shows different characteristics. Except for a similar slow rise region initially, the PDMS/PLA-7 specimen exhibits superior energy absorption efficiency including the highest efficiency peak (>0.8) and a wider efficiency stable region. The last region is a rapid decay region. The energy absorption efficiency of specimen PDMS/PLA-7 is undoubtedly the highest.

According to Equation (2), it is found that σ_{max} has a direct impact on the energy absorption efficiency. For the PDMS/PLA-3 specimen, there is a stress peak on the compressive stress–strain curves. Meanwhile, the compressive strength is much higher than the plateau stress due to the stress drop. This makes σ_{max} much higher than the stress at the subsequent strain. Therefore, the energy absorption efficiency of specimen PDMS/PLA-3 is relatively low. For the PDMS/PLA-5 specimen, the compressive strength and plateau stress are almost identical, meaning that σ_{max} is close to the plateau stress at the subsequent strain. This improves the energy absorption efficiency of specimen PDMS/PLA-5. However, the decay stage comes early due to the narrow plateau region. For the PDMS/PLA-7 specimen, a wider plateau region makes σ_{max} almost equal to the plateau stress within the plateau region. This contributes to obtaining a wider stable region and higher energy absorption efficiency. This result reflects that the strut diameter has an obvious impact on the energy absorption efficiency.

The energy absorption density of co-continuous interlocking PDMS/PLA lattice composites at the 0.6 strain is depicted in Figure 9a. At the same time, the effect of the specimen density on energy absorption is considered. Therefore, the energy absorption density W_V at the 0.6 strain and the specific energy absorption density W_m at the 0.6 strain of co-continuous interlocking PDMS/PLA lattice composites are comparatively analyzed. The specific energy absorption density W_m is defined as W_V/ρ. However, considering the effect of density, the PDMS/PLA-7 specimen still exhibits the highest specific energy absorption density. Figure 9b shows the energy absorption density and corresponding density ρ of co-continuous interlocking PDMS/PLA lattice composites. It is found that the energy absorption density and the density are still linearly correlated. The energy absorption density increases with the specimen density increasing. This result verifies that the variation of density can affect the energy absorption density of co-continuous interlocking PDMS/PLA lattice composites.

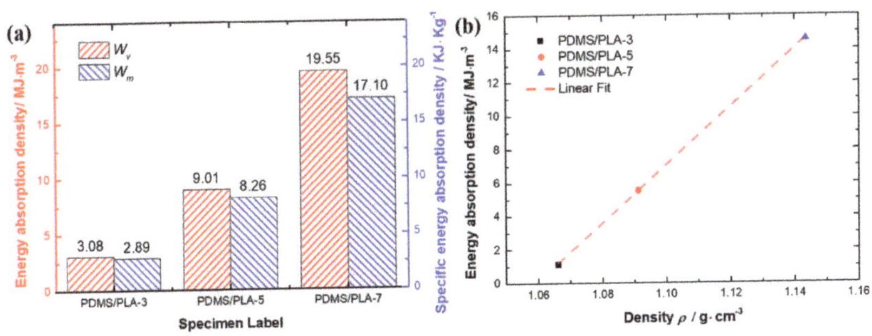

Figure 9. (**a**) The energy absorption density and the specific energy absorption density of co-continuous interlocking PDMS/PLA lattice composites. (**b**) The energy absorption density and corresponding density ρ of co-continuous interlocking PDMS/PLA lattice composites.

3.4. Effect of PLA Lattice Volume Fraction on Compressive Properties and Energy Absorption

As the co-continuous interlocking structure, the two phases, PLA and PDMS, play different roles. The PDMS fills the pores of the PLA lattice and wraps the PLA struts, playing a supporting role. Therefore, the PDMS as a filler prevents the negative effects resulting from the sudden rupture of PLA struts. Based on the aforementioned analysis, it is found that the PLA lattice acting as skeleton plays an important role in bearing the compressive load and energy absorption. In this study, changing the strut diameter actually alters the volume fraction of the PLA. With the strut diameter increasing, the volume of the PLA lattice increases. The relationships between the compressive strength, plateau stress, energy absorption density and the PLA volume fraction V_f of the co-continuous interlocking PDMS/PLA lattice composites are plotted in the Figure 10. The compressive strength σ_c and PLA volume fraction V_f are linear in logarithmic coordinates (Figure 10a). With the PLA volume fraction increasing, the compressive strength increases. The slope and intercept are 1.349 and 1.614, respectively.

Similarly, the plateau stress σ_p and PLA volume fraction V_f show a similar linear relationship in the logarithmic coordinates (Figure 10b). The slope and intercept are 2.127 and 1.914, respectively. The effect of the PLA volume fraction increasing on the plateau stress is more sensitive, meaning that the increase in the plateau stress is higher than the increase in compressive strength with the same volume fraction variation. For the energy absorption density W_m and PLA volume fraction V_f, a linear relationship is presented with a slope of 1.881 and an intercept of 1.634 (Figure 10c). The energy absorption density increases with the PLA volume fraction increasing. Therefore, the relationships between the compressive strength, plateau stress, energy absorption density and PLA volume fraction of the co-continuous interlocking PDMS/PLA lattice composites are all linear in logarithmic coordinates.

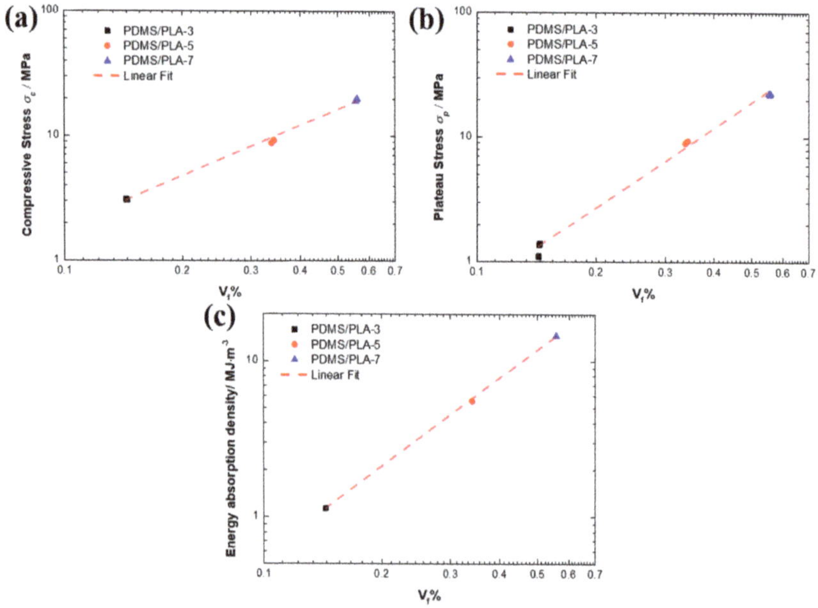

Figure 10. The relationships between (**a**) compressive strength, (**b**) plateau stress, (**c**) energy absorption density and PLA volume fraction V_f of the co-continuous interlocking PDMS/PLA lattice composites.

The PDMS/PLA lattice composites were developed based on lattice structures. For practical applications of lattice structures in energy absorption, the PDMS/PLA lattice composites can also be utilized. For instance, the PDMS/PLA lattice composites have the potential to be used in protection devices such as the crush box of automotive bumper systems due to the lightweight and energy absorption characteristics. The practical application will be examined in the ongoing study.

4. Conclusions

The conclusions are mainly summarized as follows:

- Co-continuous interlocking PDMS/PLA lattice composites were successfully manufactured using the FDM technique and infiltration method. The changes of structure parameters can be simultaneously realized using this fabrication method.
- The compressive properties of the co-continuous interlocking PDMS/PLA lattice composites can be adjusted in a narrow density range, simultaneously demonstrating a linear correlation characteristic. With the increase in the PLA strut diameter, the energy absorption density of the co-continuous interlocking PDMS/PLA lattice composites increases. Simultaneously, the energy absorption efficiency presents a higher efficiency peak and wider plateau region.
- The PLA lattice acting as skeleton plays an important role in bearing the compressive load and energy absorption. The relationships between the compressive strength, plateau stress, energy absorption density and the PLA volume fraction of co-continuous interlocking PDMS/PLA lattice composites are all linear in logarithmic coordinates. The effect of the PLA volume fraction increasing on the plateau stress is more sensitive than the compressive strength and energy absorption density.
- The PDMS/PLA lattice composites usually show superior compressive properties and energy absorption behavior compared with polymer lattice due to the introduction of

a filler. Compared with metallic or ceramic lattice composites, they usually exhibit lower compressive strength and lightweight characteristics.

Author Contributions: Conceptualization, H.W. and J.L.; Data curation, H.W., K.W. and J.L.; Formal analysis, H.W.; Funding acquisition, H.W. and X.F.; Investigation, H.W. and K.W.; Methodology, H.W. and J.L.; Project administration, X.F.; Resources, H.W. and X.F.; Validation, K.W. and J.L.; Visualization, H.W. and K.W.; Writing—original draft, H.W.; Writing—review and editing, H.W., K.W., J.L. and X.F. All authors have read and agreed to the published version of the manuscript.

Funding: This research was funded by the National Science and Technology Major Project (J2019-IV-0003-0070), the National Natural Science Foundation of China (12102335), the Aeronautical Science Foundation of China (20220041070006) and the China Postdoctoral Science Foundation (2021M702604).

Institutional Review Board Statement: Not applicable.

Informed Consent Statement: Not applicable.

Data Availability Statement: The raw data supporting the conclusions of this article will be made available by the authors upon request.

Conflicts of Interest: The authors declare that they have no known competing financial interests or personal relationships that could have appeared to influence the work reported in this paper.

References

1. Meyers, M.A.; Chen, P.; Lin, A.Y.; Seki, Y. Biological materials: Structure and mechanical properties. *Prog. Mater. Sci.* **2008**, *53*, 1–206. [CrossRef]
2. Ghazlan, A.; Ngo, T.; Tan, P.; Xie, Y.M.; Tran, P.; Donough, M. Inspiration from Nature's body armours—A review of biological and bioinspired composites. *Compos. Part B Eng.* **2021**, *205*, 108513. [CrossRef]
3. Zhang, M.; Zhao, N.; Yu, Q.; Liu, Z.; Qu, R.; Zhang, J.; Li, S.; Ren, D.; Berto, F.; Zhang, Z.; et al. On the damage tolerance of 3-D printed Mg-Ti interpenetrating-phase composites with bioinspired architectures. *Nat. Commun.* **2022**, *13*, 3247. [CrossRef]
4. Zheng, X.; Lee, H.; Weisgraber, T.H.; Shusteff, M.; DeOtte, J.; Duoss, E.B.; Kuntz, J.D.; Biener, M.M.; Ge, Q.; Jackson, J.A.; et al. Ultralight; ultrastiff mechanical metamaterials. *Science* **2014**, *344*, 1373–1377. [CrossRef]
5. Wan, T.; Liu, Y.; Zhou, C.; Chen, X.; Li, Y. Fabrication, properties, and applications of open-cell aluminum foams: A review. *J. Mater. Sci. Technol.* **2021**, *62*, 11–24. [CrossRef]
6. Mao, A.; Zhao, N.; Liang, Y.; Bai, H. Mechanically Efficient Cellular Materials Inspired by Cuttlebone. *Adv. Mater.* **2021**, *33*, 2007348. [CrossRef]
7. Li, D.; Bu, X.; Xu, Z.; Luo, Y.; Bai, H. Bioinspired Multifunctional Cellular Plastics with a Negative Poisson's Ratio for High Energy Dissipation. *Adv. Mater.* **2020**, *32*, 2001222. [CrossRef]
8. Zhang, Q.; Yang, X.; Li, P.; Huang, G.; Feng, S.; Shen, C.; Han, B.; Zhang, X.; Jin, F.; Xu, F.; et al. Bioinspired engineering of honeycomb structure—Using nature to inspire human innovation. *Prog. Mater. Sci.* **2015**, *74*, 332–400. [CrossRef]
9. Fan, J.; Zhang, L.; Wei, S.; Zhang, Z.; Choi, S.; Song, B.; Shi, Y. A review of additive manufacturing of metamaterials and developing trends. *Mater. Today* **2021**, *50*, 303–328. [CrossRef]
10. Wei, Y.; Li, H.; Yang, H.; Ma, Y.; Cheng, J.; Gao, P.; Shi, J.; Yu, B.; Lin, F. Damping and Mechanical Properties of Epoxy/316L Metallic Lattice Composites. *Materials* **2023**, *16*, 130. [CrossRef]
11. Bai, P.; Yang, X.; Shen, X.; Zhang, X.; Li, Z.; Yin, Q.; Jiang, G.; Yang, F. Sound absorption performance of the acoustic absorber fabricated by compression and microperforation of the porous metal. *Mater. Des.* **2019**, *167*, 107637. [CrossRef]
12. Cao, L.; Fu, Q.; Si, Y.; Ding, B.; Yu, J. Porous materials for sound absorption. *Compos. Commun.* **2018**, *10*, 25–35. [CrossRef]
13. Ma, P.; Wang, H.; Li, D.; Yuan, S.; Fan, X. Passive acoustic liners for aero-engine noise control: A review. *Mech. Adv. Mater. Struct.* **2024**, 1–16. [CrossRef]
14. Sharma, P.; Pandey, P.M. Morphological and mechanical characterization of topologically ordered open cell porous iron foam fabricated using 3D printing and pressureless microwave sintering. *Mater. Des.* **2018**, *160*, 442–454. [CrossRef]
15. Noronha, J.; Dash, J.; Rogers, J.; Leary, M.; Brandt, M.; Qian, M. Titanium Multi-Topology Metamaterials with Exceptional Strength. *Adv. Mater.* **2024**, 2308715. [CrossRef]
16. Wang, H.; Fu, Y.; Su, M.; Hao, H. Effect of structure design on compressive properties and energy absorption behavior of ordered porous aluminum prepared by rapid casting. *Mater. Des.* **2019**, *167*, 107631. [CrossRef]
17. Wang, H.; Fu, Y.; Su, M.M.; Hao, H. Fabrication and Compression Investigation of the Ordered Porous Aluminum with Cubic Pores. *Mater. Sci. Forum* **2018**, *933*, 97–105. [CrossRef]
18. Wang, H.; Fu, Y.; Su, M.; Hao, H. A novel method of indirect rapid prototyping to fabricate the ordered porous aluminum with controllable dimension variation and their properties. *J. Mater. Process. Technol.* **2019**, *266*, 373–380. [CrossRef]

19. Bao, H.; Li, A. Study on Quasi-Static Uniaxial Compression Properties and Constitutive Equation of Spherical Cell Porous Aluminum-Polyurethane Composites. *Materials* **2018**, *11*, 1261. [CrossRef]
20. Cheng, H.F.; Han, F.S. Compressive behavior and energy absorbing characteristic of open cell aluminum foam filled with silicate rubber. *Scr. Mater.* **2003**, *49*, 583–586. [CrossRef]
21. Wang, H.; Su, M.; Hao, H. Compressive properties and energy absorption behavior of Mg17Al12/Al ordered structure composites. *Compos. Part B Eng.* **2021**, *210*, 108688. [CrossRef]
22. Zhang, Y.; Hsieh, M.; Valdevit, L. Mechanical performance of 3D printed interpenetrating phase composites with spinodal topologies. *Compos. Struct.* **2021**, *263*, 113693. [CrossRef]
23. Wang, L.; Lau, J.; Thomas, E.L.; Boyce, M.C. Co-Continuous Composite Materials for Stiffness, Strength, and Energy Dissipation. *Adv. Mater.* **2011**, *23*, 1524–1529. [CrossRef]
24. Liu, Y.; Wang, L. Enhanced stiffness, strength and energy absorption for co-continuous composites with liquid filler. *Compos. Struct.* **2015**, *128*, 274–283. [CrossRef]
25. Al-Ketan, O.; Soliman, A.; AlQubaisi, A.M.; Al-Rub, R.K.A. Nature-Inspired Lightweight Cellular Co-Continuous Composites with Architected Periodic Gyroidal Structures. *Adv. Eng. Mater.* **2018**, *20*, 1700549. [CrossRef]
26. Wang, K.; Wang, H.; Zhang, J.; Fan, X. Mechanical behavior of interpenetrating phase composite structures based on triply periodic minimal surface lattices. *Compos. Struct.* **2024**, *337*, 118044. [CrossRef]
27. Al-Ketan, O.; Assad, M.A.; Al-Rub, R.K.A. Mechanical properties of periodic interpenetrating phase composites with novel architected microstructures. *Compos. Struct.* **2017**, *176*, 9–19. [CrossRef]
28. Al-Ketan, O.; Al-Rub, R.K.A.; Rowshan, R. Mechanical Properties of a New Type of Architected Interpenetrating Phase Composite Materials. *Adv. Mater. Technol.* **2017**, *2*, 1600235. [CrossRef]
29. Mansouri, M.R.; Montazerian, H.; Schmauder, S.; Kadkhodapour, J. 3D-printed multimaterial composites tailored for compliancy and strain recovery. *Compos. Struct.* **2018**, *184*, 11–17. [CrossRef]
30. Chen, Y.; Wang, L. Periodic co-continuous acoustic metamaterials with overlapping locally resonant and Bragg band gaps. *Appl. Phys. Lett.* **2014**, *105*, 191907. [CrossRef]
31. Liu, Y.; Tan, Q.; Lin, H.; Wang, J.; Wang, K.; Peng, Y.; Yao, S. Integrated design and additive manufacturing of lattice-filled multi-cell tubes. *Compos. Sci. Technol.* **2023**, *243*, 110252. [CrossRef]
32. Ali, N.B.; Khlif, M.; Hammami, D.; Bradai, C. Mechanical and morphological characterization of spherical cell porous structures manufactured using FDM process. *Eng. Fract. Mech.* **2019**, *216*, 106527.

materials

MDPI

Article

Carbon Nanotubes–Gr Inspired by Geckos' Setae Structure with Enhanced Tribological Properties

Jing Zhang [1], Yang Sun [1,2,*], Fengqin Shang [1], Zihan Yan [1], Jiayu Yao [1], Binghuan Chen [1] and Hangyan Shen [1,*]

1 College of Materials and Chemistry, China Jiliang University, Hangzhou 310018, China; p22050854065@cjlu.edu.cn (J.Z.); p22050854042@cjlu.edu.cn (F.S.); s23050805024@cjlu.edu.cn (Z.Y.); p23050854056@cjlu.edu.cn (J.Y.); p24050858001@cjlu.edu.cn (B.C.)

2 School of Materials Science and Engineering, Tianjin University, Tianjin 300354, China

* Correspondence: yangsun11@cjlu.edu.cn (Y.S.); shenhangyan@cjlu.edu.cn (H.S.)

Abstract: The setae structure of geckos' toes can create a strong adhesion force, allowing geckos to climb almost vertical walls. Inspired by this, carbon nanotubes–graphite (CNTs-Gr) was prepared by microwave technology, where CNTs like the setae structure grew in situ on the surface of Gr flakes. Compared to the Gr, the coefficient of friction (COF) and wear rate of CNTs-Gr decreased by 44% and 46%, reaching 0.10 and 1.18×10^{-5} $mm^3 \cdot N^{-1} \cdot m^{-1}$, respectively. Even if the load increased from 5 N to 35 N, the CNTs-Gr maintained a low and stable COF of 0.12. The excellent tribological properties were attributed to the unique setae structure of CNTs-Gr. This structure enabled the adhesion force of CNTs-Gr to the worn surface to increase threefold, improving the coverage of the lubricating film and significantly enhancing the lubricating film's pressure resistance. The gecko setae structure proposed in this article provides researchers with a new idea for designing lubricants with excellent lubrication performance and high load-bearing capacity.

Keywords: setae structure; CNTs-Gr; microwave technology; tribological properties; lubricating film

1. Introduction

Friction and wear are the main causes of energy loss and mechanical failure [1,2]. Adding lubricating materials is one of the most effective ways to reduce friction and wear, which is significant for energy conservation and prolonging mechanical life [3,4]. Common lubricating materials, such as graphite (Gr) [5–7], carbon nanotubes (CNTs) [8–10], MoS_2 [11], etc., could form a lubricating film to isolate the contact between friction pairs, which is the key to reducing friction and wear [12].

Researchers have found that the area of lubricating film on the worn surface is an important factor affecting friction and wear performance. Xiao et al. [11] studied the friction properties of Cu-MoS_2 composites and observed that expanding the coverage area of the MoS_2 film on the worn surface from 0% to 70% resulted in a decrease in the COF from 0.76 to 0.18. The reduction in the COF could be attributed to the MoS_2 lubricating film, which reduced the direct contact between the friction pairs. Its layered structure also provided excellent shear properties, further contributing to the reduction in friction. Similarly, Sun et al. [13] found that the Ag shell/Cu core structure effectively prevented an interfacial reaction between the Cu core and WS_2 during hot-press sintering. As a result, the area of WS_2 lubricating film on the worn surface of (Ag shell/Cu core)-WS_2 composites was 64% higher than that on Cu-WS_2 composites, resulting in a decrease in COF and wear rate by 66% and 79%, respectively. Therefore, increasing the area of lubricating film was an effective measure to decrease friction and wear. To increase the lubricating film area, Wei et al. [14] grafted chitosan-based copolymers onto the surface of graphene oxide, which could quickly and firmly adhere to the worn surface, significantly increasing the area of the lubricating film by 28%. Wang et al. [15] prepared silver-doped carbon quantum dots (Ag-CQDs) through the hydrothermal method and found that Ag-CQDs could adsorb onto the worn surface through chemical reactions with the substrates, resulting in a 9% increase in the area of the lubricating film to carbon quantum dots (CQDs). The lubrication performance of Ag-CQDs was significantly improved, with a reduction of 12% in COF and 8% in wear rate, respectively. The lubricating mechanism was attributed to the formation of a lubricating film composed of Ag-CQDs, multicomponent oxides, carbonates and nitrides, induced by tribochemical reactions, which provided effective protection for the worn surface and maintained a low COF and wear. Thus, enhancing the adhesion between lubricating materials and the worn surface was significant for improving the friction and wear properties. Although the grafting and hydrothermal methods used in the above works achieved success in enhancing the adhesion performance of lubricants, their preparation processes required lengthy synthesis steps and harsh reaction conditions, as well as the use of toxic catalysts and organic solvents [16,17]. Therefore, an efficient, convenient, and pollution-free method is needed to prepare lubricants with high adhesion performance.

Nature has provided excellent inspiration for designing and fabricating functional materials [18], such as shark skin swimsuits [19], imitation of lotus leaf hydrophobic films [20], and imitating spider web capture structures [21]. Gecko toes are also a common inspiration for building functional materials. Geckos can crawl lightly on glass surfaces thanks to the setae structure on their feet, as shown in Figure 1. This unique multi-stage setae structure can form good contact with the substrate surface and increase adhesion [22,23]. Typically, the gecko's adult body weight is 50 to 100 g, and a gecko's single foot can produce an adhesion force of 100 times gravity [24]. Ge et al. [25] transferred carbon nanotube arrays to flexible polymer ribbons to simulate setae structures to obtain four times the shear force of gecko feet. Therefore, constructing a gecko-like setae structure on the surface of lubricating materials improved their adhesion, and their lubrication performance was expected to be enhanced.

Figure 1. (**a**) Setae structure on a single toe of a gecko; (**b**) locally enlarged view of the setae structure. ((**a**) was the magnified image of the area within the red box).

In this study, inspired by the gecko setae structure, microwave technology was used to grow CNTs on the surface of Gr flakes, with the CNTs growing in situ on the surface of the Gr flakes, resembling the setae structure of a gecko. Microwave technology was used to utilize the microwave absorption properties of Gr to rapidly, efficiently, and without pollution grow CNTs in situ. The tribological properties of CNTs-Gr were comprehensively studied using a ball-disk friction testing machine, the microstructure of the worn surface was characterized, and the lubrication mechanism of CNTs-Gr was explored. Specifically, the CNTs-Gr adhered to the worn surface through the strong adhesive force provided by the setae structure, forming a lubricating film with a large area, which reduced friction and wear.

2. Materials and Methods

2.1. Materials

Gr powder (99% purity, 6.5 μm), ferrocene (99% purity), and Naphthalene (99% purity) power were selected as the raw materials for preparing the CNTs-Gr. Gr and ferrocene power were purchased from Shanghai Aladdin Biochemical Technology Co., Ltd (Shanghai, China). Naphthalene power was purchased from Shanghai McLean Biochemical Technology Co., Ltd (Shanghai, China). CNTs were selected as the lubricant additives and were purchased from Nanjing Xianfeng Nanomaterials Technology Co. (Nanjing, China). Polyvinylpyrrolidone (PVP, 99% purity) was selected as the dispersant in the water-based lubricant and was purchased from Shanghai McLean Biochemical Technology Co., Ltd (Shanghai, China). All reagents were analytically pure.

2.2. Preparation of CNTs-Gr

The preparation process of the CNTs-Gr is illustrated in Figure 2. Firstly, Gr (200 mg), ferrocene (100 mg), and naphthalene (100 mg) were placed into an agate mortar and ground for 10 min. Next, the mixed powder was placed into a sealed quartz bottle and microwaved at 800 W power for 1 min. Then, the microwave-treated powder was ground with ferrocene (50 mg) and naphthalene (50 mg), followed by microwave treatment again at 800 W power for 1 min. Finally, the CNTs-Gr was successfully prepared by the above scheme.

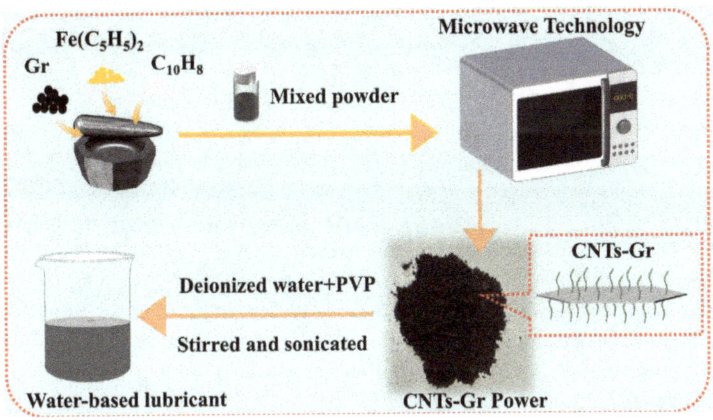

Figure 2. Schematic of the preparation process for CNTs-Gr.

2.3. Preparation of CNTs-Gr Water-Based Lubricant

Gr, CNTs, CNTs/Gr (the mechanical mixture CNTs and Gr, CNTs: Gr = 1:2), and CNTs-Gr were added as lubricant additives to the deionized water to prepare four different water-based lubricants, in which the concentration of all lubricant additives was 1.0 wt%. Then, PVP (0.1 wt%) was added as a dispersant to the water-based lubricants. Finally, all dispersions were stirred for 1 h using a magnetic stirrer and sonicated for 1 h at room temperature to obtain the uniformly mixed water-based lubricants, as shown in Figure 2.

2.4. Friction Test and Characterization

The friction and wear performance of the water-based lubricants was studied using a ball-disk friction testing machine (CSM Instrμments, Peseux, Switzerland) at room temperature. The top of the tester was a Si_3N_4 ball (Φ6 mm), and the disk was a polished stainless steel (SS) disk. During the friction process, both the SS disk and the Si_3N_4 ball were fully immersed in the prepared water-based lubricant. The entire friction experiment was conducted under submerged conditions in the water-based lubricant. Based on previous work and scientific references, the friction test was conducted under a load of 5 N at a speed of 0.5 cm/s for 15 min [26], and each experiment was repeated 3 times. After the friction test, the 3D profile of the wear track on the SS disk was determined utilizing a probe surface profiler (P-6, KLA-Tencor, Milpitas, CA, USA), which provided comprehensive data (cross-sectional profile, s) on the surface topography. The value of the wear volume (V) was calculated through V = s × l, in which l was the length of the wear track. The wear rate K of the worn surface was obtained by the following equation [27]:

$$K = V/(F \times L) \tag{1}$$

where K is the wear rate ($mm^3 \cdot N^{-1} \cdot m^{-1}$), V is the wear volume (mm^3), L is the total sliding distance (m), and F is the normal load (5 N).

The morphology and structure of the CNTs-Gr were observed using a scanning electron microscope (SEM, SU8010, Hitachi Corporation, Tokyo, Japan), which allowed for high-resolution imaging of the sample surface and the identification of surface features at the micro- and nanoscale. The SEM could measure the elemental composition using an integrated Energy-Dispersive Spectroscopy (EDS) system, providing elemental maps and qualitative/quantitative analyses of the sample composition. The SEM equipped with EDS was used to analyze the elemental composition and distribution characteristics of the lubricating film. Transmission electron microscopy (TEM, TF20, FEI Company, Hillsboro, OR, USA) was used to study the lattice structure of CNTs-Gr at resolutions as low as the nanoscale. The phase structure of CNTs-Gr was characterized using an X-ray diffractometer (XRD, X'Pert3 Powder, PANalytical Company, Almelo, The Netherlands). The XRD technique provided phase identification and crystallographic information by measuring the diffraction pattern of X-rays that were scattered by the atomic planes in the sample. The XRD could operate over a 2θ range of 5° to 90° and could measure the lattice spacing and phase composition. However, a limitation of XRD was that it primarily analyzed the crystalline structures and might not have detected amorphous or poorly crystallized regions of the sample. A 3D laser scanning microscope (KEYENCE, VK-X1050, Osaka, Japan) was utilized to observe the worn surface on the SS disk. The 3D laser microscope provided precise height measurements and surface roughness mapping at the micro- and nanoscales. A difficulty during measurement was the surface preparation, as any contamination or roughness on the SS disk could have affected the accuracy of the worn surface measurement.

3. Discussion

3.1. *Characterization of CNTs-Gr*

Figure 3 presents the microstructures of Gr and CNTs-Gr, which were synthesized through microwave technology. It can be seen in Figure 3a that the Gr showed a typical sheet-like structure, the surface of which was very smooth. According to the process shown in Figure 2, when the mixture of Gr, ferrocene, and naphthalene was heated by microwave technology, the ferrocene decomposed into nanocatalyst particles and anchored onto the Gr surface, as shown in Figure 3b. Then, these nanocatalyst particles absorbed the carbon source provided by naphthalene decomposition and promoted the growth of CNTs through the tip growth mechanism [28]. As a result, CNTs-Gr was formed, as shown in Figure 3c, d, in which numerous CNTs grew along the smooth surface of the Gr flakes, creating a structure resembling gecko setae. Figure 3e,f are enlarged images of the top structure of CNTs in CNTs-Gr (Figure 3d), showing that the nearly circular nanocatalyst particles were completely wrapped in the walls of CNTs. The nanocatalytic particles continuously catalyzed the formation of CNTs, which were pushed upwards by the grown CNTs [28]. Furthermore, a vertically upward structure of CNTs was formed on the surface of the Gr. In addition, image analysis software (ImageJ 1.x) was used to analyze over 100 CNTs in the CNTs-Gr, where the average diameter of the CNTs was 101.01 nm, and their diameter distribution is shown in Figure 3g. The average length of the CNTs was 0.83 μm, and their distribution density on the Gr surface was 7.57/μm^2. Figure 3h shows that the CNTs-Gr was composed of C and Fe elements, in which the C element content accounted for 95.42 wt%. The atomic percentage was analyzed, with C accounting for 98.98 at%, and Fe accounting for 1.02 at%. Among them, the Fe element was mainly provided by the catalyst particles from the decomposed ferrocene [28].

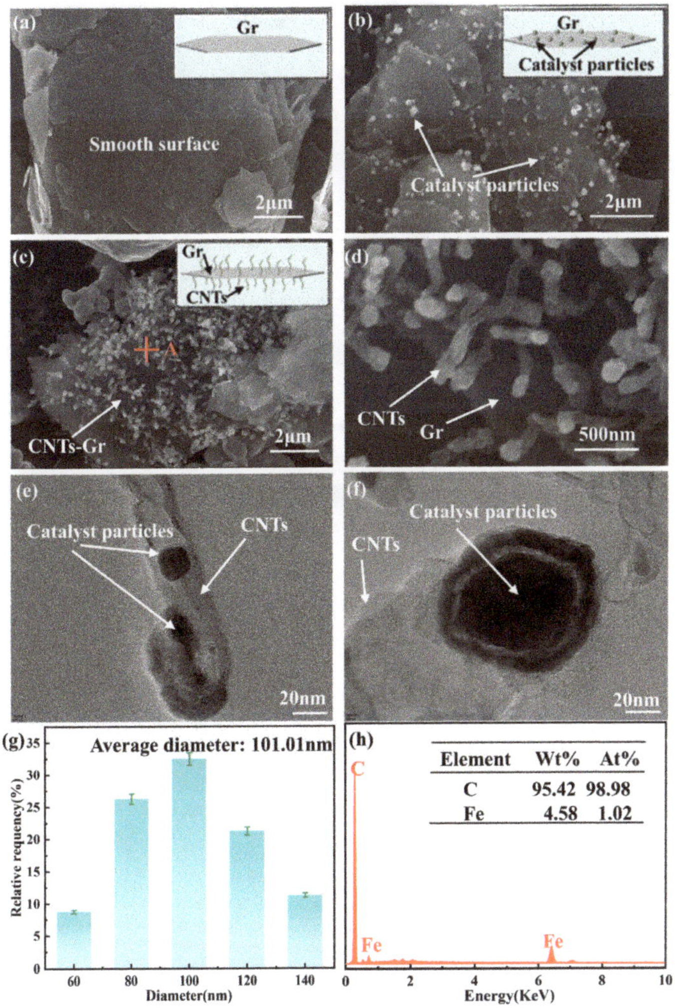

Figure 3. SEM images of (**a**) Gr and (**b**–**d**) CNTs-Gr; TEM images of (**e**,**f**) CNTs in CNTs-Gr; (**g**) diameter distribution of CNTs; and (**h**) EDS spectra of CNTs-Gr. (The letters and symbols in (**c**) represent the location of the EDS).

Figure 4 presents the XRD patterns of Gr and CNTs-Gr. The characteristic peaks of Gr and CNTs-Gr at 26.6°, 42.4°, and 54.6° could be attributed to the (002), (100), and (004) crystal planes of Gr (JCPDS No. 898487). In addition, unique peaks were observed in the XRD pattern of CNTs-Gr, with the peak at 43.7° corresponding to the (111) plane of carbide iron (Fe_3C) and the peak at 44.6° corresponding to the (110) plane of iron (Fe) [29]. The results indicated that the nanocatalysts in Figure 3e were Fe and Fe_3C. Prominent peaks in the XRD spectrum showed that the CNTs-Gr prepared by microwave technology were composed of CNTs, Gr, and catalyst particles (Fe_3C, Fe).

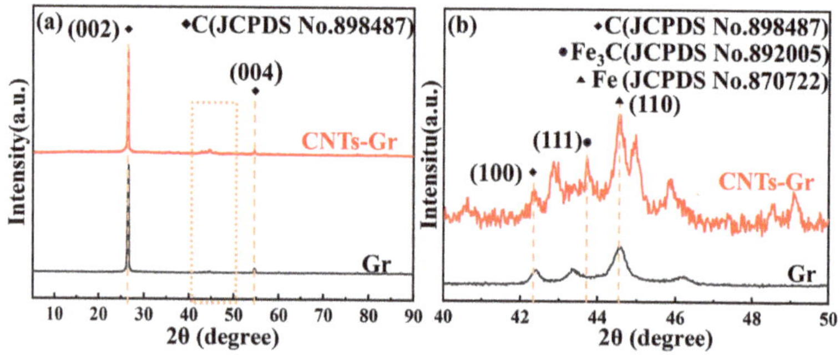

Figure 4. (**a**) XRD pattern of Gr and CNTs-Gr and (**b**) local enlarged pattern.

3.2. *Tribological Behavior of CNTs-Gr and Characterization of the Worn Surface*

The COF and wear rate of the CNTs-Gr and control samples (Gr, CNTs, CNT/Gr) are presented in Figure 5. It can be observed from Figure 5 that the COF and wear rate of the pure water were the highest, which were 0.62 and 2.82×10^{-5} $mm^3 \cdot N^{-1} \cdot m^{-1}$, respectively. After adding PVP (0.1 wt%) to the pure water, the COF and wear rate were similar to those of pure water, indicating that PVP had little effect on the lubrication performance of pure water. However, when Gr and CNTs were added into water, the lubrication performance of pure water could be significantly improved. The COFs of the Gr and CNTs decreased to 0.17 and 0.32, respectively, and the wear rates decreased to 1.68×10^{-5} $mm^3 \cdot N^{-1} \cdot m^{-1}$ and 2.35×10^{-5} $mm^3 \cdot N^{-1} \cdot m^{-1}$, respectively. More interestingly, when CNTs-Gr was used as an additive, the COF and wear rate were the lowest, at 0.10 and 1.18×10^{-5} $mm^3 \cdot N^{-1} \cdot m^{-1}$, respectively. Compared to the Gr, there was a reduction of 44.1% in the COF and 46.1% in the wear rate. Therefore, the CNTs-Gr exhibited excellent tribological performance and possessed a lower COF and wear rate.

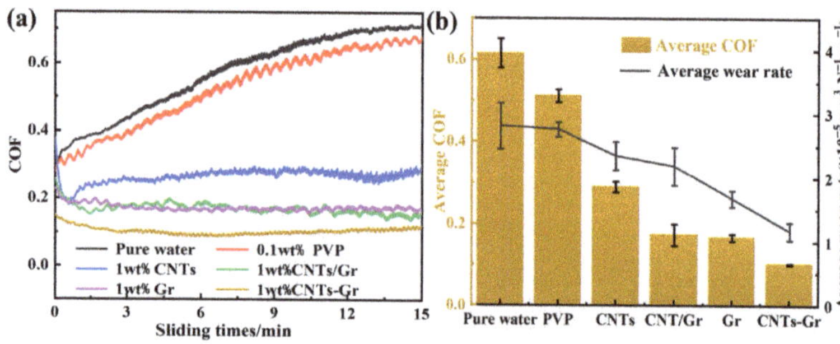

Figure 5. (**a**) The COF curves of pure water, PVP, CNTs, CNTs/Gr, Gr, and CNTs-Gr and (**b**) average COF and wear rate.

In order to obtain the wear condition of Si_3N_4 balls, the wear surface was characterized using SEM, as shown in Figure 6. The wear surface diameters were ordered from small to large as CNTs-Gr, Gr, CNTs/Gr, and CNTs. The wear surface diameter of CNTs-Gr was 367.9 μm, while that of CNTs was 503.3 μm, as shown in Figure 6a, d. The wear rate of the Si_3N_4 ball was calculated according to the model [30] shown in Figure 6e, and the result is shown in Figure 6f. Compared to Gr and CNTs, the wear rate of the CNTs-Gr decreased by 71.2% and 31.6 %, respectively. When CNTs-Gr was used as a lubricant, both Si_3N_4 balls and SS disks (as shown in Figure 5b) exhibited the lowest wear rate, confirming that CNTs-Gr had superior lubrication and anti-wear properties compared to Gr and CNT.

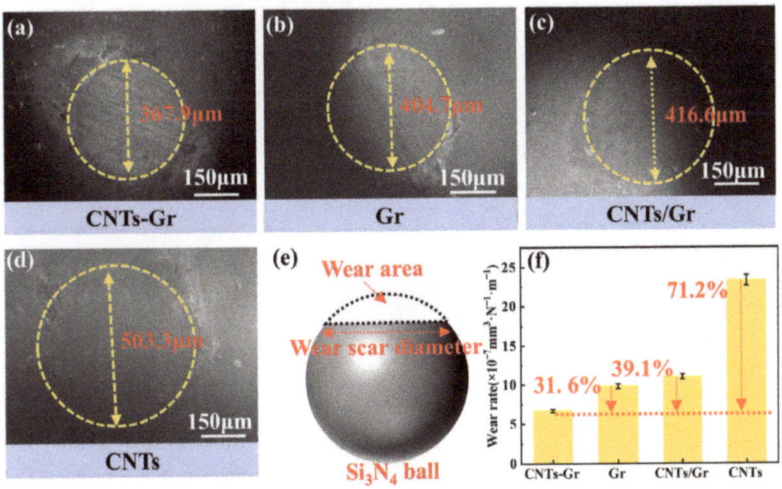

Figure 6. SEM images of the worn surfaces of the Si_3N_4 ball: (**a**) CNTs-Gr, (**b**) Gr, (**c**) CNTs/Gr, and (**d**) CNTs lubricant. (**e**) Calculation model and (**f**) wear rate of Si_3N_4 ball.

To investigate the lubrication mechanism, a detailed analysis was conducted on the worn surface of the SS disk. Figure 7(a1–d2) show the worn surfaces and their magnified images under lubrication of CNTs-Gr, Gr, CNTs/Gr, and CNTs, respectively. The minimum diameter of the worn surface with CNTs-Gr was 142 μm, representing reductions of 25.7% and 61.55% compared to Gr and CNTs, respectively. By magnifying the worn surface, it was found that the dark black area on the worn surface was the lubricating film. As shown in Figure 7(a2), a continuous lubricating film was present on the worn surface of the SS disk that was lubricated with CNTs-Gr, and under high-magnification SEM, CNTs were observed on the surface of the lubricating film. This indicated that the setae structure of CNTs-Gr could effectively adhere to the worn surface, forming a lubricating film with a large area. It was precisely this lubricating film with a large area that led to the reductions in the COF and wear rate [11,31]. In Figure 7(b2), it can be seen that Gr formed a partial lubricating film on the worn surface, and the worn surface exhibits deep grooves. In Figure 7(d2), it can be seen that the worn surface under the CNT lubricant shows almost no lubricating film and a large amount of wear debris, produced by friction. At the same time, aggregates of CNTs were also observed on the worn surface, which could lead to an increase in the COF and wear rate [32,33]. Therefore, CNTs-Gr was able to form a continuous and complete lubricating film structure on the worn surface, leading to reductions in COF and wear.

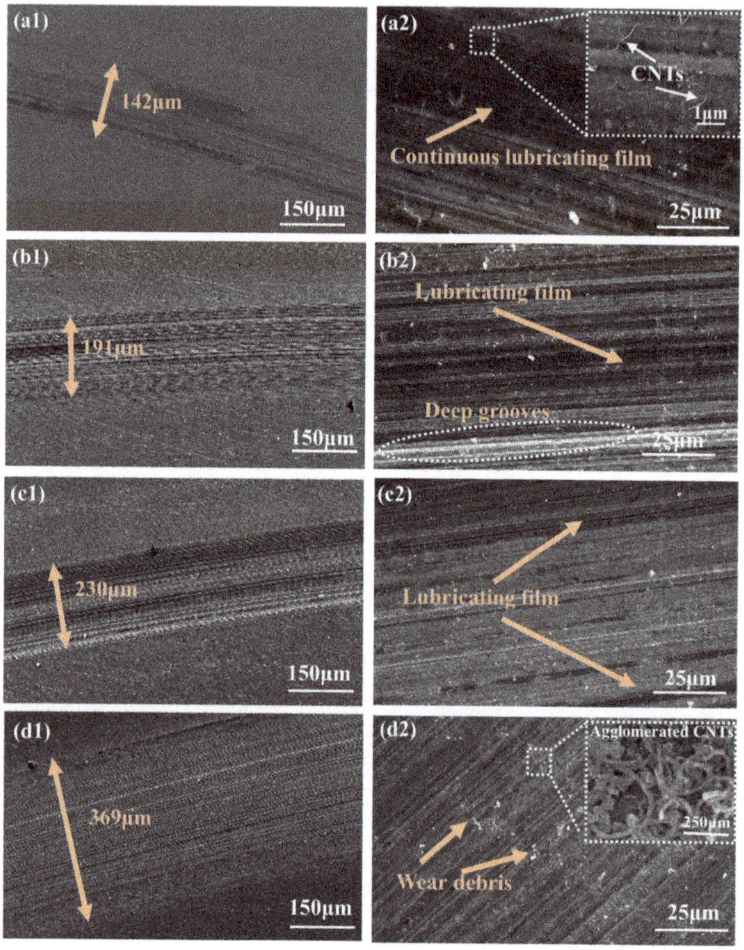

Figure 7. (**a1**,**a2**), (**b1**,**b2**), (**c1**,**c2**), and (**d1**,**d2**), respectively, depict the worn surfaces of SS disks when using CNTs-Gr, Gr, CNTs/Gr, and CNTs lubricant.

A 3D laser scanning microscope was further used to evaluate the worn surfaces of the SS disk. Figure 8 presents the 3D profiles and cross-sectional wear scar curves of the worn surfaces under the four lubricant additives of CNTs-Gr, Gr, CNTs/Gr, and CNTs. As shown in Figure 8(d1,d2), the 3D profiles of CNTs were severely serrated, and the height of the worn surface burr was very irregular. In Figure 8(c1,c2), with the addition of CNTs/Gr, the sawtooth phenomenon is weakened and the irregularity of the burr height is reduced. Figure 8(a1,a2) shows that the serrated shape of the 3D profiles of CNTs-Gr was the least obvious, which indicated that the roughness of its worn surface was significantly reduced. This was due to the formation of a continuous lubricating film structure on the surface by CNTs-Gr, as shown in Figure 7(a2). Furthermore, the surface roughness parameter Sa, calculated from the 3D profile, revealed that CNTs-Gr had the smallest Sa (0.337 μm), while CNTs exhibited the largest Sa (0.841 μm). The increase in the surface roughness of the worn surface led to an increase and fluctuation in the COF and an increase in the wear rate [31].

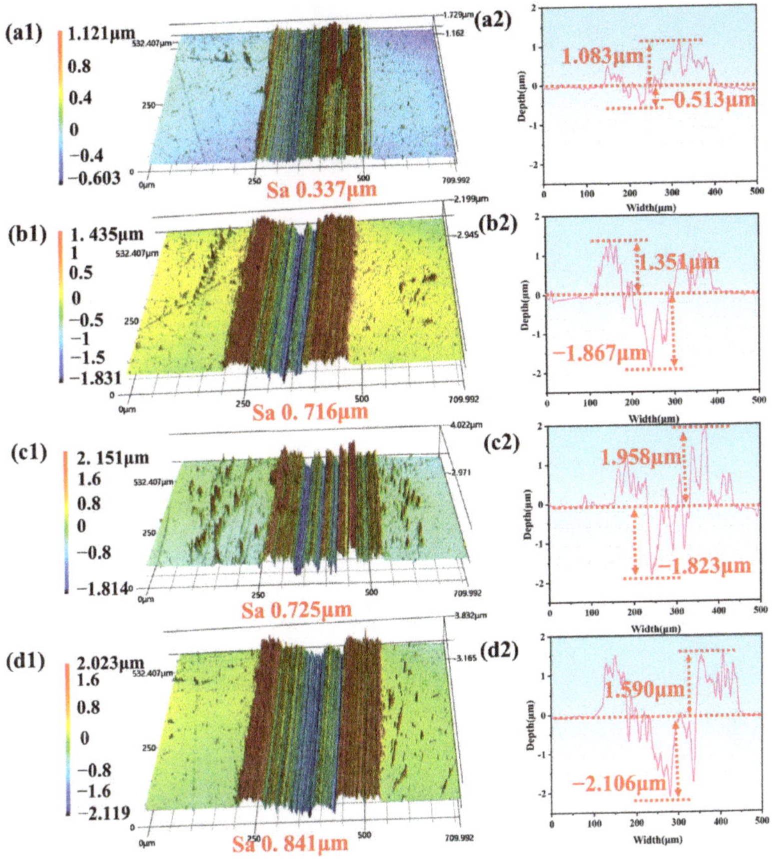

Figure 8. Three-dimensional profiles and cross-sectional wear scar curves of SS with different lubricant additives. (**a1**,**a2**) CNTs-Gr, (**b1**,**b2**) Gr, (**c1**,**c2**) CNTs/Gr, and (**d1**,**d2**) CNTs.

To compare the tribological properties of Gr and CNTs-Gr, the sliding time was increased to 80 min. Figure 9a shows the COF curves of the Gr and CNTs-Gr. Figure 9b shows the average COF and standard deviation (SD) every 20 min. The smooth COF curves represent a more stable tribological performance of the lubricating materials, and the SD was introduced to evaluate the stability [34,35]. In Figure 9b, it can be seen that at the sliding times of 0–20 min, the COF of CNTs-Gr was 0.10, with an SD of 15×10^{-3}. As the sliding times increased, the COF and SD of CNTs-Gr gradually decreased to 0.08 and 0.05×10^{-3} (60–80 min), respectively. At the sliding times of 0–20 min, the COF of Gr was 0.17, with an SD of 20×10^{-3}. As the sliding times increased, the COF and SD of Gr decreased to 0.15 and 2×10^{-3} (60–80 min), respectively. Compared to Gr, the CNTs-Gr exhibited lower COFs and SDs. The stable tribological performance of CNTs-Gr stemmed from CNTs-Gr adhering to the worn surface of the disk, thereby generating a lubricating film [36], which reduced the COF and SD [37].

Figure 10 shows the SEM and EDS images of the worn surface of the SS disk for CNTs-Gr (a–d) and Gr (e–h). Since the lubrication film was mainly composed of C, the coverage of the lubrication film on the worn surface could be confirmed by the C content, and the range rate of the lubricating film was statistically analyzed using image analysis software (ImageJ). The coverage of the lubricating film increased with the increase in the sliding times. At 20 min, the worn surfaces of CNTs-Gr and Gr were covered by 25.2% and

15.5% lubricating films, respectively, as shown in Figure 10(a2,e2). However, at 80 min, the coverage area of the CNTs-Gr lubricating film reached 56.8%, which was 121.9% higher than that under the action of Gr. The EDS results indicated that CNTs-Gr effectively adhered to the worn surface, forming a lubricating film with a large area, which reduced the COF and wear rate.

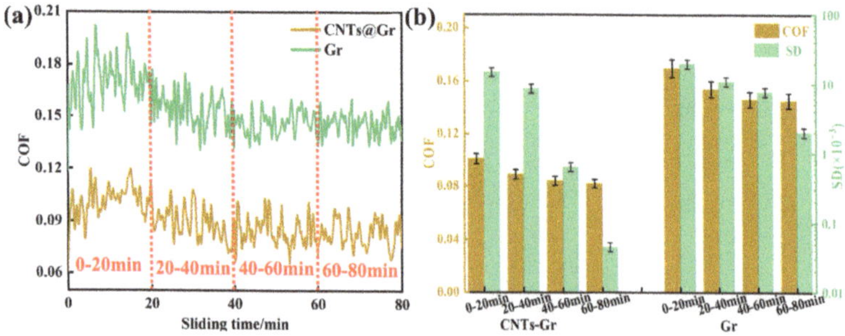

Figure 9. (**a**) The COF curves of Gr and CNTs-Gr; (**b**) average COF and SD every 20 min.

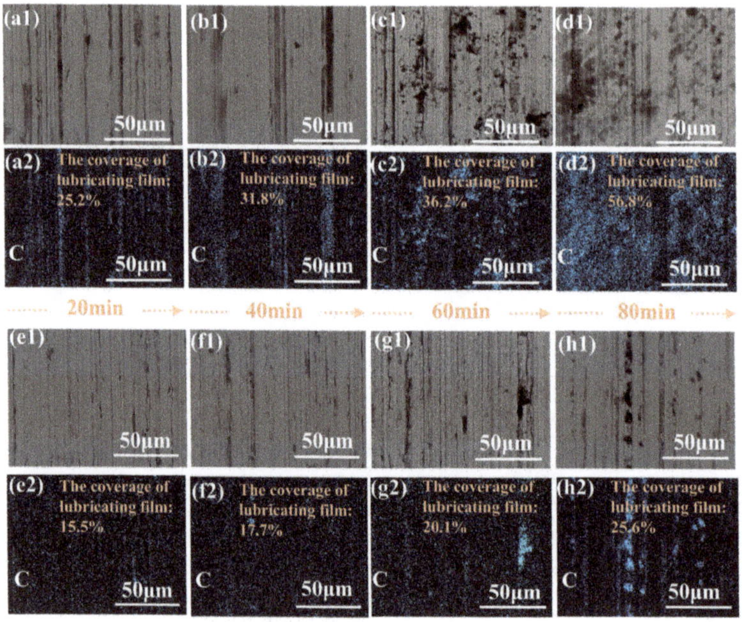

Figure 10. SEM and EDS images of worn surfaces on SS disks, lubricated with (**a1–d2**) CNTs-Gr and (**e1–h2**) Gr, within the sliding times 20, 40, 60, and 80 min.

In Figure 11a, b, it can be seen that as the load increased from 5 N to 35 N, the COF of Gr rose by 41.2%. However, the COF of CNTs-Gr remained relatively stable under varying loads, with an increase of only 20.0% as the load went from 5 N to 35 N. This phenomenon was attributed to the setae structure of CNTs-Gr, which could still firmly adhere to the worn surface even under a 35 N load, ensuring that the lubricating film remained stable despite the increase in load.

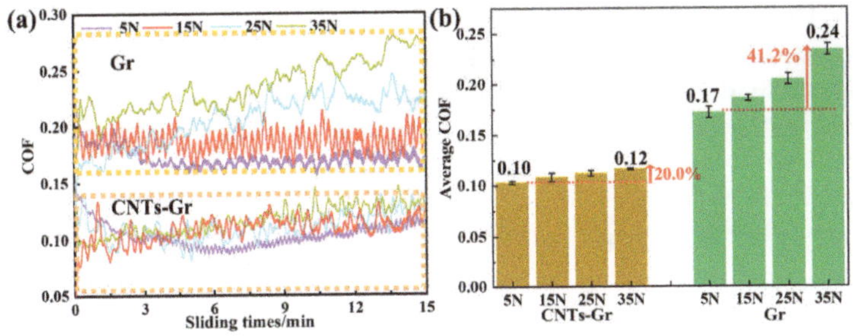

Figure 11. (**a**) The COF curves of Gr and CNTs-Gr under 5–35 N loads and (**b**) the average COF of Gr and CNTs-Gr under 5–35 N loads.

Figure 12 shows the worn surfaces under Gr and CNTs-Gr lubrication at 5 N and 35 N loads. As shown in Figure 12a,c, the dark areas in the SEM images represented the regions covered by the lubricating film, and the range rate of the lubricating film was statistically analyzed using image analysis software (ImageJ). Under a 5 N load, the coverage of the lubricating film on the worn surface under CNTs-Gr lubrication was 25.5%. When the load increased to 35 N, the coverage of the lubricating film remained at 23.3%. Therefore, as the load increased, the COF of CNTs-Gr increased by only 20.0%, as shown in Figure 11b. In contrast, under a 5 N load, the coverage of the lubricating film on the worn surface under Gr lubrication was 14.4%, and when the load increased to 35 N, the coverage of the lubricating film decreased to only 7.8%, a reduction of 45.8%. As shown in Figure 12d, under Gr lubrication at 35 N, deep grooves appeared on the worn surface. This indicated that as the load increased, the lubricating film structure formed by Gr on the worn surface was largely destroyed, leading to a 41.2% increase in the COF, as shown in Figure 11b.

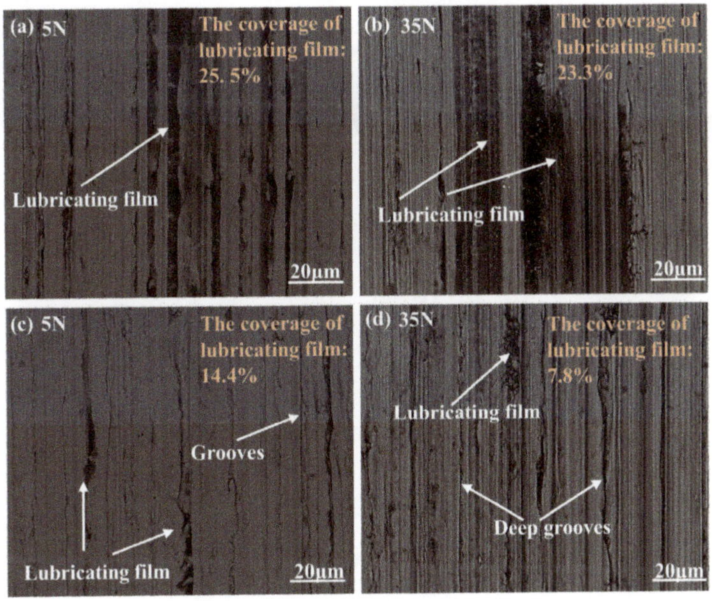

Figure 12. The SEM images of worn surfaces of CNTs-Gr lubricants under (**a**) 5 N and (**b**) 35 N and Gr lubricant under (**c**) 5 N and (**d**) 35 N.

3.3. Analysis of Lubrication Mechanism

The above results indicated that CNTs-Gr could adsorb onto the worn surface and form a continuous lubricating film due to its setae structure. The adhesion force primarily originated from molecular interactions, whose strength was proportional to the contact area [38]. If the CNTs-Gr and SS substrates were regarded as two planes that were primarily attracted by van der Waals forces, the adhesion force at the contact interface could be approximated by the following equation [39]:

$$F = HA/\left(6\pi d^3\right) \tag{2}$$

where H is the Hamiltonian constant, A is the contact area, and d is the distance between the surfaces. Taking a typical H value of 0.4×10^{-19} J and d of about 1 nm [39], the contact area between the CNTs-Gr and the substrate ($A_{CNTs\text{-}Gr}$) was the sum of the Gr's contact area (A_{Gr}) and the CNTs' contact area (A_{CNTs}), as shown in the following equation:

$$A_{CNTs-Gr} = A_{Gr} + A_{CNTs} \tag{3}$$

The enhanced adhesion of CNTs-Gr was attributed to CNTs filling the voids of the substrate and increasing the contact area. Taking Gr as a plate and CNTs as a cylinder, it was assumed that the CNTs-Gr were in full contact with the substrate. The diameter of the CNTs was 101.01 nm, and the length was 0.83 μm, as determined by the image analysis software (ImageJ). The distribution density of CNTs on the Gr surface was 7.57/μm^2. After the calculation, the contact area between 1 μm^2 of CNTs-Gr and the substrate was 3.05 μm^2. Using Equation (2), the adhesion forces of Gr and CNTs-Gr with the substrate were 2.12×10^{-8} N/μm^2 and 6.47×10^{-8} N/μm^2, respectively. Therefore, CNTs-Gr formed a continuous lubricating film structure, attributed to the strong adhesion force provided by its setae structure. The friction mechanism schematic of CNTs-Gr is shown in Figure 13. During the friction test, CNTs-Gr entered the friction area and increased the contact area with the worn surface through the setae structure, which improved the adhesion and formed the lubricating film. As the sliding proceeded, CNTs-Gr continued to adsorb on the worn surface, increasing the area of the lubricating film, which resulted in a gradual decrease in the COF to a stable value. Consequently, CNTs-Gr could avoid direct contact with the friction pair, providing efficient lubrication and anti-wear properties.

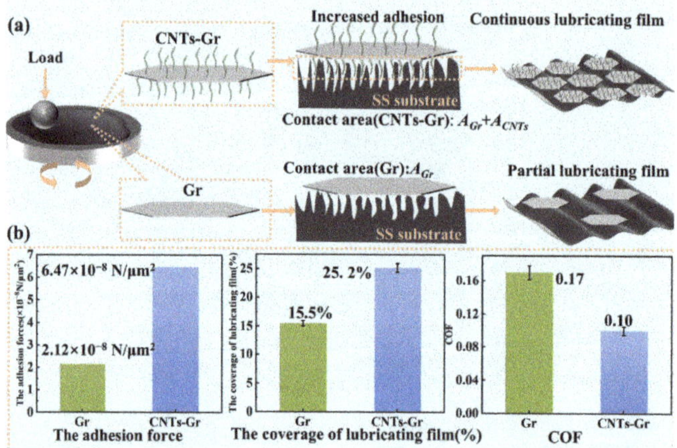

Figure 13. (**a**) Friction mechanism of CNTs-Gr additives; (**b**) The adhesion force, the coverage of lubricating film, and COF of Gr and CNTs-Gr.

4. Conclusions

In this study, a new technology was proposed for the rapid synthesis of CNTs-Gr, and its microstructure and chemical properties were characterized in detail. In addition, the tribological properties and lubrication mechanism of the CNTs-Gr were systematically investigated. The prepared CNTs-Gr, due to its setae structure, exhibited excellent adhesion and could effectively adsorb onto the worn surface, forming a lubricating film with a large area. This structure significantly reduced the COF and wear, aligning with the anticipated performance improvements. The conclusions were as follows:

(1) CNTs-Gr was successfully prepared within 2 min using microwave technology. The CNTs in CNTs-Gr grew vertically and uniformly on the surface of Gr flakes, forming a gecko setae-like structure.

(2) The unique setae structure of CNTs-Gr led to an excellent lubrication performance. The COF and wear rate of the CNTs-Gr were 0.10 and 1.18×10^{-5} $mm^3 \cdot N^{-1} \cdot m^{-1}$, respectively. Compared to Gr, the COF and wear rate decreased by 44.1% and 46.1%, respectively. Even when the load was increased to 35 N, the COF of the CNTs-Gr remained stable at 0.12.

(3) The setae structure of CNTs-Gr tightly contacted the worn surface, generating strong adhesion (6.47×10^{-8} $N/\mu m^2$) and promoting the formation of a lubricating film during the friction test. After 90 min of friction test, the lubricating film coverage reached up to 56.8%. Even under high loads, the lubricating film maintained good coverage.

(4) CNTs-Gr, with its excellent lubrication performance, could replace Gr as a lubricant in parts such as engines, brake systems, and wheel bearings. The preparation process only required a few minutes of microwave technology, thus not increasing the cost of use. Since the CNTs were grown in situ on the surface of Gr, this avoided the potential harm of nanoparticles to humans and the environment.

Author Contributions: Conceptualization, B.C. and H.S.; Methodology, J.Z., F.S., and Z.Y.; Investigation, J.Z. and Z.Y.; Validation, J.Y.; Resources, H.S.; data curation, J.Z.; Writing—original draft preparation, J.Z.; Writing—review and editing, Y.S. and H.S.; Supervision, F.S. and H.S.; Funding acquisition, Y.S. All authors have read and agreed to the published version of the manuscript.

Funding: The research was supported by Zhejiang Provincial Natural Science Foundation (LQ24E040001) and Zhejiang Provincial College Students Science and Technology Innovation Activity Plan Project (2024R409043).

Institutional Review Board Statement: Not applicable.

Informed Consent Statement: Not applicable.

Data Availability Statement: The original contributions presented in this study are included in the article. Further inquiries can be directed to the corresponding authors.

Acknowledgments: We would like to thank Weibo Guo, Gansheng Zhang, and Zhejiang Hengcheng Cemented Carbide Co., Ltd. for not only providing financial support but also making significant contributions to the methodology and conceptualization of this paper.

Conflicts of Interest: The authors declare no conflicts of interest.

References

1. Holmberg, K.; Erdemir, A. The impact of tribology on energy use and CO2 emission globally and in combustion engine and electric cars. *Tribol. Int.* **2019**, *135*, 389–396. [CrossRef]
2. He, F.; Xie, G.; Luo, J. Electrical bearing failures in electric vehicles. *Friction* **2020**, *8*, 4–28. [CrossRef]
3. Luo, J.; Zhou, X. Superlubricitive engineering—Future industry nearly getting rid of wear and frictional energy consumption. *Friction* **2020**, *8*, 643–665. [CrossRef]
4. Wang, K.; Wu, H.; Wang, H.; Liu, Y.; Yang, L.; Zhao, L. Tribological properties of novel palygorskite nanoplatelets used as oil-based lubricant additives. *Friction* **2020**, *9*, 332–343. [CrossRef]

5. Bai, L.; Ge, Y.; Zhu, L.; Chen, Y.; Yi, M. Preparation and properties of copper-plated expanded graphite/copper composites. *Tribol. Int.* **2021**, *161*, 107094. [CrossRef]
6. Paul, G.; Shit, S.; Hirani, H.; Kuila, T.; Murmu, N.C. Tribological behavior of dodecylamine functionalized graphene nanosheets dispersed engine oil nanolubricants. *Tribol. Int.* **2019**, *131*, 605–619. [CrossRef]
7. Wu, H.; Zhang, H.; Gao, A.; Gong, L.; Ji, Y.; Zeng, S.; Li, S.; Wang, X. Friction and wear performance of aluminum-based self-lubricating materials derived from the 3D printed graphite skeletons with different morphologies and orientations. *Tribol. Int.* **2024**, *195*, 109614. [CrossRef]
8. Shan, Z.; Jia, X.; Tian, R.; Yang, J.; Fan, H.; Cellulose-armored, H.S. "CNTs-soft metal" hybrid nanomaterials to improving the friction-reduction and anti-wear performance of bio-lubricants. *Tribol. Int.* **2024**, *191*, 109085. [CrossRef]
9. Chandran, A.; Unnikrishnan, N.; Jayaraj, M.; John, R.E.; George, J. *Recent Advances in Graphene and Graphene-Based Technologies*; IOP Publishing: Bristol, UK, 2023.
10. Barhoum, A.; Jeevanandam, J.; Danquah, M.K. *Fundamentals of Bionanomaterials*; Elsevier: Amsterdam, The Netherlands, 2022.
11. Xiao, J.; Wu, Y.; Zhang, W.; Chen, J.; Zhang, C. Friction of metal-matrix self-lubricating composites: Relationships among lubricant content, lubricating film coverage, and friction coefficient. *Friction* **2019**, *8*, 517–530. [CrossRef]
12. Zhao, J.; Huang, Y.; He, Y.; Shi, Y. Nanolubricant additives: A review. *Friction* **2020**, *9*, 891–917. [CrossRef]
13. Sun, Y.; Fang, Y.; Shang, F.; Zhang, J.; Liu, W.; Shen, H.; Guo, B. Enhancing the tribological performance of Cu-WS2 composites with Ag-shell/Cu-core structure. *Powder Technol.* **2023**, *422*, 118453. [CrossRef]
14. Wei, Q.; Fu, T.; Yue, Q.; Liu, H.; Ma, S.; Cai, M.; Zhou, F. Graphene oxide/brush-like polysaccharide copolymer nanohybrids as eco-friendly additives for water-based lubrication. *Tribol. Int.* **2021**, *157*, 106895. [CrossRef]
15. Wang, J.; Li, X.; Deng, Y.; Chen, S.; Liang, W.; Zhang, L.; Wei, X.; Gao, S.; Wan, Y. Carbon quantum dots doped with silver as lubricating oil additive for enhancing tribological performance at various temperatures. *Appl. Surf. Sci.* **2022**, *599*, 154029. [CrossRef]
16. Wen, P.; Lei, Y.; Li, W.; Fan, M. Two-dimension layered nanomaterial as lubricant additives: Covalent organic frameworks beyond oxide graphene and reduced oxide graphene. *Tribol. Int.* **2020**, *143*, 106051. [CrossRef]
17. Sun, J.; Li, A.; Su, F. Excellent Lubricating Ability of Functionalization Graphene Dispersed in Perfluoropolyether for Titanium Alloy. *ACS Appl. Nano Mater.* **2019**, *2*, 1391–1401. [CrossRef]
18. He, Q.; Tang, T.; Zeng, Y.; Iradukunda, N.; Bethers, B.; Li, X.; Yang, Y. Review on 3D Printing of Bioinspired Structures for Surface/Interface Applications. *Adv. Funct. Mater.* **2023**, *34*, 2309323. [CrossRef]
19. Dai, W.; Alkahtani, M.; Hemmer, P.R.; Liang, H. Drag-reduction of 3D printed shark-skin-like surfaces. *Friction* **2018**, *7*, 603–612. [CrossRef]
20. Wang, D.; Huang, J.; Guo, Z. Tomato-lotus inspired edible superhydrophobic artificial lotus leaf. *Chem. Eng. J.* **2020**, *400*, 125883. [CrossRef]
21. Wang, W.; Yu, B.; Sun, W.; Jiang, C.; Lu, L. Bioinspired nanostructured spiderweb for high-efficiency capturing and killing of bacteria. *Sci. China Mater.* **2021**, *65*, 518–526. [CrossRef]
22. Song, Y.; Wang, Z.; Li, Y.; Dai, Z. Electrostatic attraction caused by triboelectrification in climbing geckos. *Friction* **2020**, *10*, 44–53. [CrossRef]
23. Russell, A.P.; Stark, A.Y.; Higham, T.E. The Integrative Biology of Gecko Adhesion: Historical Review, Current Understanding, and Grand Challenges. *Integr Comp Biol* **2019**, *59*, 101–116. [CrossRef] [PubMed]
24. Autumn, K.; Liang, Y.A.; Hsieh, S.T.; Zesch, W.; Chan, W.P.; Kenny, T.W.; Fearing, R.; Full, R.J. Full. Adhesive force of a single gecko foot-hair. *Nature* **2000**, *405*, 681–685. [CrossRef] [PubMed]
25. Ge, L.; Sethi, S.; Ci, L.; Ajayan, P.M.; Dhinojwala, A. Carbon nanotube-based synthetic gecko tapes. *Proc. Natl. Acad. Sci. USA* **2007**, *104*, 10792–10795. [CrossRef]
26. Benamor, A.; Hadji, Y.; Kota, S.; Chiker, N.; Liu, L.; Haddad, A.; Sahraoui, T.; Ying, G.; Hadji, M. Friction and wear characteristics of the nanolaminated ternary transition metal boride: Mn2AlB2. *Wear* **2022**, *492–493*, 204232. [CrossRef]
27. Ren, K.; Yu, G.; Zhang, Z.; Wu, W.; Tian, P.; Chhattal, M.; Gong, Z.; Li, Y.; Zhang, J. Self-organized transfer film-induced ultralow friction of Graphene/MoWS4 heterostructure nanocomposite. *Appl. Surf. Sci.* **2022**, *572*, 151443. [CrossRef]
28. Bajpai, R.; Wagner, H.D. Fast growth of carbon nanotubes using a microwave oven. *Carbon* **2015**, *82*, 327–336. [CrossRef]
29. Nie, H.; Cui, M.; Russell, T.P. A route to rapid carbon nanotube growth. *Chem. Commun. (Camb)* **2013**, *49*, 5159–5161. [CrossRef]
30. Wang, S.; Ren, G.; Li, W.; Wang, B.; Wei, F.; Liang, Z.; Chen, D. A green modification technology of carbon nanotubes toward enhancing the tribological properties of aqueous-based lubricants. *Tribol. Int.* **2023**, *180*, 108268. [CrossRef]
31. Zhang, X.; Zhang, K.; Kang, X.; Zhang, L. Friction maps and wear maps of Ag/MoS2/WS2 nanocomposite with different sliding speed and normal force. *Tribol. Int.* **2021**, *164*, 107228. [CrossRef]
32. Moghadam, A.D.; Omrani, E.; Menezes, P.L.; Rohatgi, P.K. Mechanical and tribological properties of self-lubricating metal matrix nanocomposites reinforced by carbon nanotubes (CNTs) and graphene–A review. *Compos. Part B Eng.* **2015**, *77*, 402–420. [CrossRef]

33. Bukvić, M.; Gajević, S.; Skulić, A.; Savić, S.; Ašonja, A.; Stojanović, B. Tribological Application of Nanocomposite Additives in Industrial Oils. *Lubricants* **2023**, *12*, 6. [CrossRef]
34. Cao, Z.; Xia, Y.; Liu, L.; Feng, X. Study on the conductive and tribological properties of copper sliding electrical contacts lubricated by ionic liquids. *Tribol. Int.* **2019**, *130*, 27–35. [CrossRef]
35. Grandin, M.; Wiklund, U. Wear phenomena and tribofilm formation of copper/copper-graphite sliding electrical contact materials. *Wear* **2018**, *398–399*, 227–235. [CrossRef]
36. Rabaso, P.; Ville, F.; Dassenoy, F.; Diaby, M.; Afanasiev, P.; Cavoret, J.; Vacher, B.; Le Mogne, T. Boundary lubrication: Influence of the size and structure of inorganic fullerene-like MoS2 nanoparticles on friction and wear reduction. *Wear* **2014**, *320*, 161–178. [CrossRef]
37. Xiao, H.; Liu, S. 2D nanomaterials as lubricant additive: A review. *Mater. Des.* **2017**, *135*, 319–332. [CrossRef]
38. Yang, K.; Wang, L.; Zou, X.; Wang, H.; Liang, C.; Zhang, D.; Wang, L.-N. Modeling bacterial adhesion on the nanopatterned surface by varying contact area. *J. Mater. Sci. Technol.* **2024**, *196*, 137–147. [CrossRef]
39. Yang, L.; Zou, M.; Wu, S.; Xu, W.; Wu, H.; Cao, A. Graphene Oxide Glue-Electrode for Fabrication of Vertical, Elastic, Conductive Columns. *ACS Nano* **2017**, *11*, 2944–2951. [CrossRef]

materials

Article

Vibration Characteristics Analysis of Boring Bar with Tunable Dynamic Vibration Absorber

Yanqi Guan [1,2,3,*], Guangbin Yu [4], Qingming Hu [1,2,3], Donghui Xu [1,2,3], Jiao Xu [5] and Pavel Lushchyk [4]

1 School of Mechanical and Electronic Engineering, Qiqihar University, Qiqihar 161006, China; huqingming1267@126.com (Q.H.); 15946212473@139.com (D.X.)
2 The Engineering Technology Research Center for Precision Manufacturing Equipment and Industrial Perception of Heilongjiang Province, Qiqihar University, Qiqihar 161006, China
3 The Collaborative Innovation Center for Intelligent Manufacturing Equipment Industrialization, Qiqihar University, Qiqihar 161006, China
4 School of Mechatronics Engineering, Harbin Institute of Technology, Harbin 150001, China; yu_ccna@163.com (G.Y.); lushchykpavel@163.com (P.L.)
5 HE Harbin Power Plant Valve Company Limited, Harbin 150066, China; xujiao_06@sina.com
* Correspondence: guanyq12345@163.com

Abstract: In deep-hole boring processes, boring bars with a large length-to-diameter ratio are typically employed. However, excessive overhang significantly reduces the boring bar's stiffness, inducing vibrational effects that severely degrade machining precision and surface quality. To address this, the research objective is to suppress vibrations using a tunable-parameter boring bar. This paper proposes a novel Tunable Dynamic Vibration Absorber (TDVA) boring bar and designs its fundamental parameters. Based on the derived dynamic model, the vibration characteristics of the proposed boring bar are analyzed, revealing the variation in damping performance under different excitation frequencies. By establishing the relationship between TDVA stiffness, damping, and the axial compression of rubber bushings, optimal parameter combinations can be precisely identified for specific excitation frequencies. Ultimately, adjusting the TDVA's axial compression displacement (0.1–0.5 mm) significantly expands the effective machining frequency range compared to conventional designs while maintaining operational reliability. This study proposes a novel Tunable Dynamic Vibration Absorber (TDVA) that innovatively integrates axial compression to achieve coupled stiffness and damping adjustments, addressing the rigidity–adaptability trade-off in deep-hole boring tools.

Keywords: deep-hole boring; vibration absorber; axial compression; boring bar

1. Introduction

In deep-hole boring, insufficient stiffness due to excessive length-to-diameter ratios in boring bars often induces severe vibrations. These vibrations not only deteriorate workpiece surface quality [1,2], but may also trigger chatter, leading to tool breakage or machining failure [3]. Consequently, suppressing boring bar vibrations has become a critical challenge for improving deep-hole machining precision. Current vibration control technologies for boring bars are categorized into passive and active methods [4,5].

Early researchers developed passive vibration suppression techniques by applying various damping principles to boring bar design. Typical examples include boring bars with special geometric structures, impact damping particles, friction energy dissipation structures [6–9], and embedded dynamic vibration absorbers (DVAs). Passive methods achieve vibration control through structural optimization or additional damping devices, offering high reliability and low cost. Recent advancements focus on variable-stiffness DVAs. For instance, Lie Li et al. [10–12] proposed a DVA embedded with rubber bushing supports, where axial compression adjusts stiffness, significantly influencing damping effects. Iklodi Z [13] proposed a methodology combining time–domain simulations and hybrid periodic orbit continuation techniques to investigate the dynamic behavior of displacement-constrained tuned mass dampers (TMDs) in boring processes, addressing the deterioration of damping performance and inherent chatter instability risks caused by such constraints. Houck [14] suppressed resonance by tuning the natural frequency matching between the boring bar and tool holder. Additionally, Haizhao Shi et al. [15] proposed an equivalent linearization method, which showed that by optimizing the stiffness, mass, and damping of the shock absorber, the vibration peak of the boring bar can be effectively reduced, and the vibration absorption effect can be improved. L. Rubio and Miguélez [16, 17] focused on optimizing the parameters of passive DVA, with the optimization criterion maximizing the minimum value of the stability lobe diagram. The calculation results indicate a significant improvement in stability performance, and the parameter tuning efficiency was improved by modifying the formula.

With the advancement of innovative materials and control technologies, various active vibration-damped boring bars have been developed. Typical active solutions include electromagnetic variable-damping boring bars, particle-damped boring bars, magnetostrictive actuator-driven boring bars, electrorheological fluid-actuated boring bars, magnetorheological fluid-controlled boring bars, and piezoelectric ceramic-driven boring bars.

Active vibration-damping technologies rely on innovative materials and closed-loop control for dynamic adjustment. Representative methodologies include a variety of approaches. Liu Qiang et al. [18] designed an electromagnetic variable-damping boring bar that dynamically regulates magnetic damping forces by adjusting coil voltage. Taha Gokulu [19] expanded the application of zero-order harmonic methods in chatter analysis by integrating multi-insert rotation and time-varying stiffness design. Jiyuan Tian [20] optimized particle damping parameters via discrete element simulations, demonstrating that tungsten steel particles (Φ0.5 mm, 90% filling rate) enhance damping performance by 70%. GUO et al. [21] developed a tunable particle damper (TPD) with frequency-adaptive stiffness. Ganesan Ramu et al. [22] proposed a vertical multi-cell hybrid particle damping system, achieving higher resonance gaps, reduced displacement, and minimal surface roughness. Lawranc et al. [23] utilized passive constrained layer damping (CLD) technology, optimizing combinations of tool substrate materials (copper/aluminum/brass) and elastomer layers (nitrile rubber/polyurethane) to suppress vibrations and extend tool life. In the field of brilliant damping, C.V. Biju et al. [24] pioneered a semi-active boring bar using magnetorheological fluid, enabling dynamic damping matching through electromagnetic regulation. Fan Chen et al. [25] implemented a magneto-actuated H∞ control system that significantly increased dynamic stiffness, resulting in chatter-free material removal rates. Yamada K et al. [26] proposed a hybrid piezoelectric–LR circuit vibration suppression method, improving energy dissipation efficiency through circuit parameter optimization and validating its industrial applicability.

While passive methods remain favored in engineering due to structural simplicity, their fixed parameters limit adaptability to varying machining conditions. Conversely, active damping solutions face challenges such as complex sensor–actuator integration and reliability concerns despite their dynamic tunability. To address these trade-offs, this study introduces a Tunable Dynamic Vibration Absorber (TDVA) boring bar. A quantitative model is established to correlate TDVA stiffness, damping, and axial compression, revealing their coupling mechanisms. This work provides a vibration control solution for deep-hole boring that balances adjustability and operational reliability.

2. Theoretical Model of the Boring Bar

In order to study the vibration reduction performance of the boring bar, it is essential to analyze its damping mechanism. The damped boring bar consists of a main body and a TDVA. The TDVA is embedded within the boring bar structure. According to the dynamic vibration absorber (DVA) theory, the TDVA serves as the critical component for vibration suppression. Installing the absorber near the tooltip of large length-to-diameter ratio boring bars can effectively mitigate external excitation-induced vibrations. During design, it is necessary to balance the overall stiffness of the boring bar with optimal vibration absorption performance. To maintain sufficient structural rigidity, the absorber's volume must be minimized while ensuring effective damping functionality.

2.1. Vibration Model of Boring Bar

The boring process is illustrated in Figure 1a. The boring bar moves along the x-axis while the workpiece rotates about its central axis. The primary cutting force acts in the F_R direction. The mathematical model of the boring bar is simplified as a rigidly clamped vibration system, as shown in Figure 1b, where M_1 represents the equivalent mass of the boring bar, K_1 denotes its equivalent stiffness, F_0 is the excitation force amplitude, m_2 and k_2 are the mass and stiffness of the TDVA, c is the damping coefficient, and x_1 and x_2 are displacements of the boring bar and TDVA, respectively. The system is further simplified to a two-degree-of-freedom model, as in Figure 2.

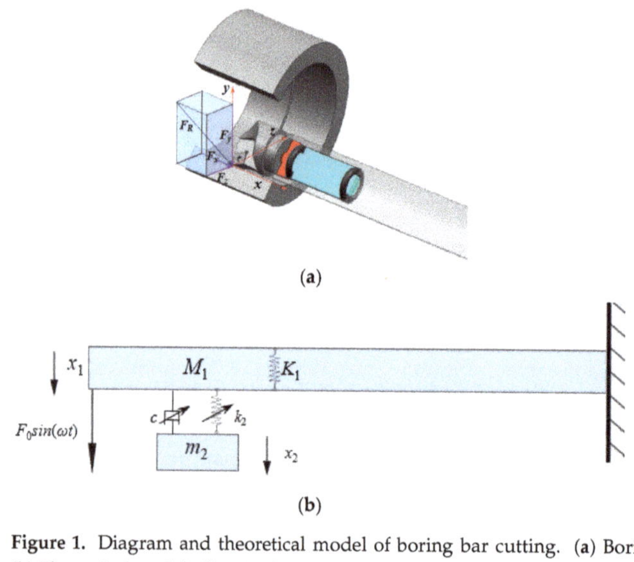

Figure 1. Diagram and theoretical model of boring bar cutting. (**a**) Boring schematic diagram; (**b**) Theoretical model of boring bar.

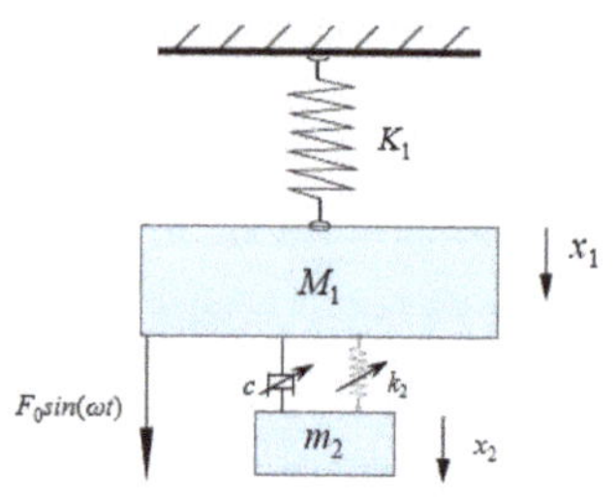

Figure 2. Two-degree-of-freedom vibration model.

When the boring bar is subjected to practical external excitation forces, the magnitude of the external loads undergoes periodic variation. The maximum amplitude of the external excitation force acting on the boring bar is defined as F_0. In the characteristic vibration analysis, load variation serves as the primary research focus. The nonlinear effects of tool–workpiece contact interactions are neglected in this study, with emphasis placed solely on the time-dependent load component. The external load is defined as a harmonically varying periodic excitation force, expressed mathematically as [27,28]

$$\begin{cases} M_1\ddot{x}_1 + c(\dot{x}_1 - \dot{x}_2) + k_1x_1 + k_2(x_1 - x_2) = F_0\sin(\omega t) \\ m_2\ddot{x}_2 + c(\dot{x}_2 - \dot{x}_1) + k_2(x_2 - x_1) = 0 \end{cases} \quad (1)$$

When Equation (1) is converted into matrix form, the following is obtained:

$$\begin{bmatrix} M_1 & 0 \\ 0 & m_2 \end{bmatrix}\begin{bmatrix} \ddot{x}_1 \\ \ddot{x}_2 \end{bmatrix} + \begin{bmatrix} c & -c \\ -c & c \end{bmatrix}\begin{bmatrix} \dot{x}_1 \\ \dot{x}_2 \end{bmatrix} + \begin{bmatrix} k_1 + k_2 & -k_2 \\ -k_2 & k_2 \end{bmatrix}\begin{bmatrix} x_1 \\ x_2 \end{bmatrix} = \begin{bmatrix} F_0\sin(\omega t) \\ 0 \end{bmatrix} \quad (2)$$

where M_1 is the equivalent mass of the boring bar, K_1 denotes its equivalent stiffness, F_0 is the excitation force amplitude, m_2 and k_2 are the mass and stiffness of the TDVA, c is the damping coefficient, x_1 is the displacement of the boring bar, x_2 is the displacement of TDVA.

Considering that the external excitation is a harmonic force, the steady-state solution in complex form can be expressed as

$$\begin{bmatrix} x_1 \\ x_2 \end{bmatrix} = \begin{bmatrix} A_1 \\ A_2 \end{bmatrix} e^{j\omega t} \quad (3)$$

Substituting the above equation into Equation (1), the maximum amplitude at the boring bar tip during cutting can be obtained as

$$A_1 = \frac{k_2 - m_2\omega^2 + j\omega c}{(K_1 + k_2 - M_1\omega^2 + j\omega c)(k_2 - m_2\omega^2 + j\omega c) - (k_2 + j\omega c)^2}F_0 \quad (4)$$

By reorganizing Equation (4), the amplitude ratio can be derived as

$$A_1 = \frac{(k_2 - m_2\omega^2)^2 + (\omega c)^2}{[(K_1 - M_1\omega^2)(k_2 - m_2\omega^2) - m_2k_2\omega^2]^2 + [K_1 - (M_1 + m_2)\omega^2]^2(\omega c)^2}F_0 \quad (5)$$

By rearranging the above equation, the amplitude ratio is derived as

$$\frac{|A_1|}{A_{st}} = \sqrt{\frac{(\lambda^2 - \gamma^2)^2 + (2\gamma\lambda\zeta)^2}{[(1 - \lambda^2)(\gamma^2 - \lambda^2) - \mu\gamma^2\lambda^2]^2 + [1 - (1 + \mu)\lambda^2]^2(2\gamma\lambda\zeta)^2}} \quad (6)$$

where $\omega_n = \sqrt{k_2/m_2}$ is the natural frequency of the vibration absorber, $\Omega_n = \sqrt{K_1/M_1}$ is the natural frequency of the boring bar, $\mu = m_2/M_1$ is the mass ratio between the boring bar and the absorber, $\gamma = \omega_n/\Omega_n$ is the frequency ratio of the absorber to the boring bar, $\lambda = \omega/\Omega_n$ is the forced vibration frequency ratio, $\zeta = c/2m_2\omega_n$ is damping ratio, $A_{st} = F_0/K_1$ is static displacement under static load.

According to the analysis of Equation (6), a larger mass ratio μ results in a more pronounced amplitude ratio response, indicating that a higher mass ratio helps reduce system vibration amplitudes and improves damping performance. When the natural frequency ratio γ approaches resonance conditions, the system exhibits a higher amplitude ratio. Appropriately adjusting γ can achieve ideal vibration suppression effects. Since

γ is influenced by the absorber's natural frequency ω_n and the boring bar's natural frequency ω_n, the latter is typically fixed. By designing the absorber's stiffness k_2, γ can be adjusted, demonstrating that the amplitude ratio is significantly affected by k_2. Additionally, the damping ratio ζ critically impacts the amplitude ratio. Modifying ζ alters the vibration amplitude and enhances system stability within specific frequency ranges.

In summary, the amplitude ratio can be optimized through absorber design, with key parameters including the absorber's mass m_2, stiffness k_2, and damping ratio ζ. To maximize damping performance while maintaining the boring bar's overall rigidity, m_2 should be maximized, and k_2 should be tunable. The effects of k_2 and ζ on the amplitude ratio are complex and interdependent. Proper adjustment of these parameters enables effective vibration control.

2.2. Design of Variable Parameter Boring Bar Structure

The proposed boring bar structure innovatively integrates a boring head, a main body, and a variable-stiffness damping vibration absorber (TDVA). A built-in cavity is designed within the front rigid–weak zone of the boring bar to house the TDVA. The TDVA employs a tungsten-based high-density alloy mass block to achieve an optimal mass ratio μ (mass block-to-boring bar ratio), enabling high-efficiency vibration damping while ensuring the primary boring bar's rigidity. The TDVA comprises an axial compression block, dual rubber bushings, and a core mass block, forming a novel coupled mechanism: the rubber bushings encapsulate the mass block and establish radial contact constraints with the boring bar's inner wall. The left bushing interfaces with the axial compression block, while the right bushing contacts the inner wall.

Based on a bolt-threaded hole transmission mechanism, rotating the bolt drives the axial compression block to generate axial displacement, compressing the dual rubber bushings to modulate radial stiffness and damping properties simultaneously. The specific structure of TDVA boring bar is shown in Figure 3. The nonlinear compressive deformation of the rubber material inherently adjusts stiffness parameters, while its viscoelastic properties enable dynamic adaptation of damping values. This stiffness–damping synergistic regulation mechanism overcomes the limitations of traditional single-parameter tuning in damped boring bars, achieving precise vibration suppression across diverse machining conditions. Existing studies predominantly focus on independent stiffness or damping adjustment strategies, particularly for rubber bushings that exhibit both stiffness-tuning and time-varying damping characteristics. However, research gaps remain in understanding the coupled effects of dual parameters. This study systematically elucidates the regulatory mechanism of stiffness–damping interactions on boring bar vibrations.

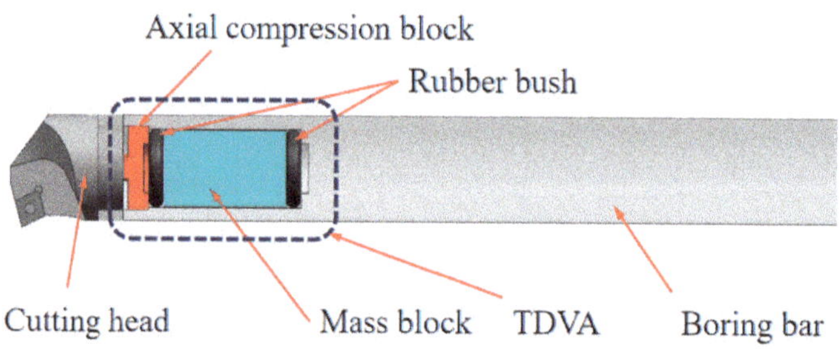

Figure 3. Structure of the boring bar with a TDVA.

3. Analysis of Vibration Characteristics of Boring Bar

For the boring bar designed in Section 2.2, key parameters, including equivalent stiffness, equivalent mass, and vibration damping ratio, are considered, as listed in Table 1.

Through rational adjustments to the variable stiffness and damping of the TDVA, the vibration characteristics of the boring bar can be significantly enhanced, thereby improving the precision and efficiency of the machining process.

Table 1. Parameters of the boring bar with a variable-stiffness TDVA.

	Equivalent Stiffness of the Boring Bar, K_1 (10^6 N/m)	Equivalent Mass of the Boring Bar, M (kg)	Equivalent Mass of the TDVA, m_2 (kg)	Damping Ratio of the TDVA, ζ	Equivalent Stiffness of the TDVA, K (10^6 N/m)
TDVA boring bar	2.2057	1.12	0.59	variable	variable
Boring bar	2.2134	1.216			

3.1. Influence of TDVA Stiffness on Boring Bar Vibration

Based on the parameters listed in Table 1, the boring bar parameters were substituted into Equation (6) to establish a vibration response model [29], yielding the TDVA stiffness-dependent vibration response curves shown in Figure 4. The horizontal axis represents the excitation frequency (0–500 Hz), while the vertical axis denotes the dimensionless amplitude ratio. The black curve corresponds to a conventional boring bar (single-degree-of-freedom system), and the colored curves represent TDVA-integrated boring bars with varying stiffness values k_2. An amplitude ratio below one indicates acceptable vibration levels for stable machining. The analysis reveals that the conventional boring bar exhibits a single intense resonance peak near 215 Hz, which aligns with its natural frequency, confirming resonance as the primary cause of machining failure. In contrast, the TDVA boring bar demonstrates typical two-degree-of-freedom system characteristics, with dual resonance peaks in all response curves: the first peak (90–190 Hz Hz) corresponds to the dominant mode of the boring bar body while the second peak (220–330 Hz) reflects the dynamic behavior of the TDVA subsystem. As k_2 increases from 3.0×10^5 N/m to 11.0×10^5 N/m, both resonance frequencies shift rightward, accompanied by reduced peak amplitudes, indicating that higher stiffness effectively broadens the stable machining frequency band.

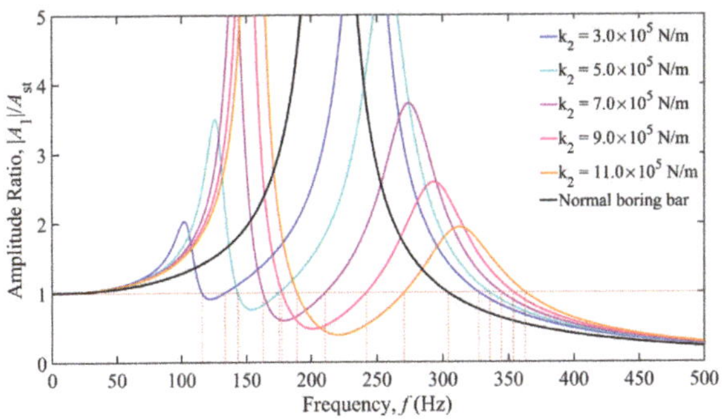

Figure 4. Amplitude ratio of the boring bar under different stiffnesses of the TDVA.

Further analysis of the influence of stiffness adjustment on vibration suppression performance reveals that when $k_2 = 3.0 \times 10^5$ N/m (blue curve), the amplitude ratio remains below one within the 115.64–133.04 Hz frequency range, yielding a stable bandwidth of 17.4 Hz. Increasing k_2 to $k_2 = 5.0 \times 10^5$ N/m (green curve) extends the stable region to 142.96–174.91 Hz (31.95 Hz bandwidth), representing a 165% improvement. The intersections of the red horizontal line Z = 1 with each curve define the critical machining frequency, which shifts toward higher frequencies as k_2 increases. At $k_2 = 11.0 \times 10^5$ N/m, the amplitude ratio remains below one in the 188.42–270.34 Hz range (81.92 Hz bandwidth);

however, the system requires 362.11 Hz to enter the unconditional stability zone at higher frequencies, where its performance is inferior to that of the conventional boring bar.

Comparing the dynamic responses of the two boring bar types reveals that the TDVA technology significantly enhances vibration suppression performance. Within the 115.64–270.34 Hz range, adjusting the TDVA stiffness satisfies machining requirements. However, in the 303.78–327.06 Hz frequency band, the conventional boring bar meets machining conditions, whereas the TDVA boring bar underperforms. By modulating the TDVA stiffness, the effective operational frequency coverage is expanded from above 303.78 Hz (conventional boring bar) to 115.64–270.34 Hz and above 327.06 Hz, achieving a 167% increase in effective frequency coverage. This provides a theoretical foundation for parameter adaptation in complex machining scenarios.

To intuitively illustrate the impact of TDVA stiffness on the amplitude ratio, a three-dimensional surface plot of the amplitude ratio under the coupled influence of frequency and TDVA stiffness is depicted in Figure 5a. The vibration characteristics of the boring bar can be classified into three distinct zones: Area A, characterized by a dual-ridge structure corresponding to the coupled resonance bands of the boring bar body (160 $\pm$ 15 Hz) and the TDVA (245 $\pm$ 20 Hz). At the ridge peaks, the amplitude ratio exceeds a value of five, requiring strict avoidance to prevent chatter. Adjustable Area C: Exhibits a saddle-shaped valley feature originating from the phase cancellation effect between stiffness and mass ratios. Within the 100–285 Hz range, adjusting the stiffness ensures an amplitude ratio below a value of one, establishing a stable cutting window with a bandwidth of 175 Hz. Area C occupies the geometric region bounded by nodal points C, D, E and the characteristic red response curve, as illustrated in Figure 5b. Area B (>343 Hz): Dominated by system inertia, the amplitude ratio decays exponentially with increasing frequency. Beyond 367 Hz, the amplitude ratio remains below a value of one across the entire frequency range, satisfying machining requirements without further adjustments. Area B is confined to the right-lateral domain demarcated by reference points F, G, and the red line red, as depicted in Figure 5b. Notably, in the 320–343 Hz transition band, modifying the TDVA stiffness reduces the amplitude ratio, validating the optimization capability of stiffness tuning for edge frequency bands.

In machining processes, the selection of cutting parameters is directly linked to the distribution characteristics of the dominant excitation frequency [30]. As shown in Figure 6, when the excitation frequency is divided into six characteristic intervals, the stiffness adaptation strategies for the TDVA exhibit significant differences. For the low-frequency band of 10–60 Hz (Figure 6a), the system operates in a stiffness-insensitive region where the amplitude ratio consistently exceeds the chatter threshold, necessitating priority optimization of the cutting path or spindle speed reduction to avoid this frequency range. When the excitation frequency increases to 70–290 Hz (Figure 6b–f), the vibration suppression effectiveness of stiffness adjustment gradually becomes prominent. Particularly within the 150–285 Hz range, maintaining the TDVA stiffness above 4.2×10^5 N/m establishes a stable machining window with a bandwidth of 135 Hz. Notably, the 270–285 Hz band requires high-stiffness configurations to suppress secondary resonance modes, while the 285–320 Hz range is classified as an absolute vibration-prohibited zone, requiring adjustments to feed rate combined with the workpiece's vibration resistance characteristics to mitigate chatter risks.

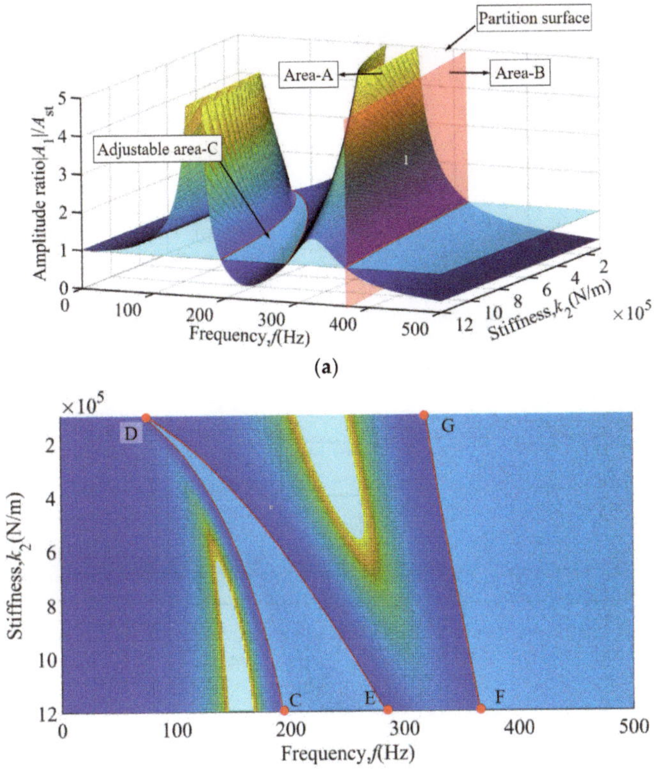

Figure 5. The amplitude ratio of the three-dimensional surface under the coupling effect of frequency and TDVA stiffness. (**a**) Macroscopic 3D surface diagram; (**b**) Top view of 3D curved surface.

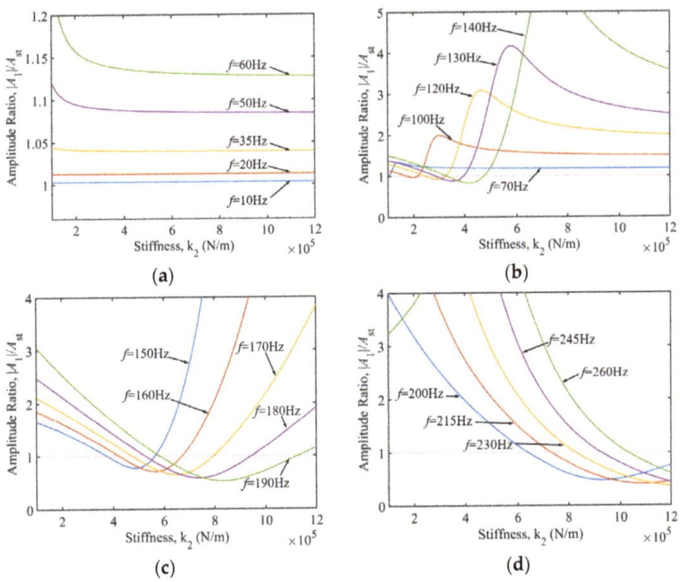

Figure 6. *Cont.*

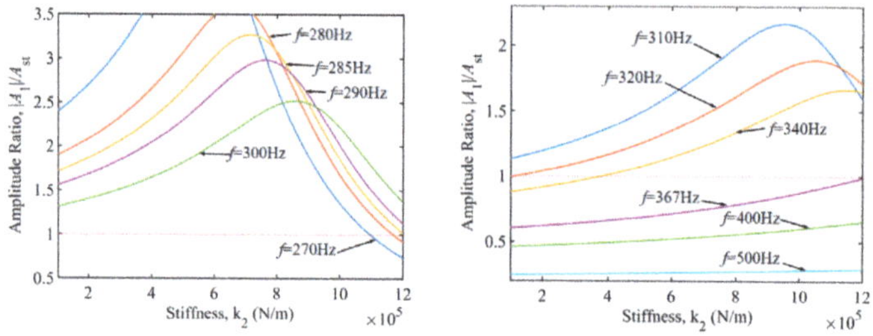

Figure 6. Control characteristics of TDVA stiffness in different frequency bands on the amplitude ratio of boring bar. (**a**) Excitation frequency 10–60 Hz; (**b**) Excitation frequency 70–140 Hz; (**c**) Excitation frequency 150–190 Hz; (**d**) Excitation frequency 200–260 Hz; (**e**) Excitation frequency 270–300 Hz; (**f**) Excitation frequency 310–500 Hz.

Further analysis of high-frequency conditions from Figure 6f reveals that within the 320–367 Hz transition band, the amplitude ratio exhibits nonlinear dependence on stiffness values, requiring dynamic adjustment of k_2 based on frequency gradients to achieve vibration suppression. When the frequency exceeds 367 Hz, the system enters an inertial stability zone, where the amplitude ratio autonomously converges below a value of one, independent of stiffness, and machining stability is ensured by maintaining baseline stiffness. This conclusively demonstrates that precise regulation of TDVA stiffness parameters is the core strategy for suppressing boring bar vibrations.

3.2. *Influence of TDVA Damping on Boring Bar Vibration*

Based on the parameters in Table 1, the damping-dependent vibration response curves of the TDVA constructed via Equation (6) are shown in Figure 7. The horizontal axis represents the excitation frequency (0–500 Hz), and the vertical axis denotes the dimensionless amplitude ratio. The black curve corresponds to the conventional boring bar, while the colored curves represent the TDVA-integrated boring bar with varying damping ratios ζ. The analysis reveals that when $\zeta < 0.4$, the system exhibits typical two-degree-of-freedom characteristics, with dual resonance peaks in the response curves. However, for $\zeta > 0.4$, the dual peaks gradually merge into a single-peak structure due to high damping suppressing vibrations in the TDVA subsystem, weakening modal coupling and significantly increasing peak amplitudes with rising ζ. The red horizontal line at an amplitude ratio of one defines the allowable vibration threshold. Its intersections with the curves mark critical stability frequency boundaries. When $0 \leq \zeta \leq 0.1$, the system establishes a stable machining interval within the 149–198 Hz band. As ζ increases to 0.4, the stable window shifts toward higher frequencies (250–367 Hz), demonstrating the directional effect of the damping ratio on frequency–domain regulation.

A detailed comparison of response patterns under varying damping ratios reveals that under low-damping conditions, the system achieves excellent vibration suppression in the low-frequency range of 148.14–198.31 Hz but is prone to inducing secondary resonance at higher frequencies. Conversely, high damping significantly suppresses high-frequency vibrations at the cost of reduced low-frequency stability. Notably, when the excitation frequency exceeds 342.49 Hz, the amplitude ratio remains below a value of one for all ζ, validating the system's inherent stability in high-frequency regions. In comparison, while the conventional boring bar retains basic machining capability above 303.78 Hz, its fixed damping structure cannot dynamically adapt to frequency-varying conditions, resulting in amplitude ratio peaks exceeding a value of five in the mid-frequency band (190–230 Hz). This highlights the technical superiority of the TDVA's tunable damping mechanism.

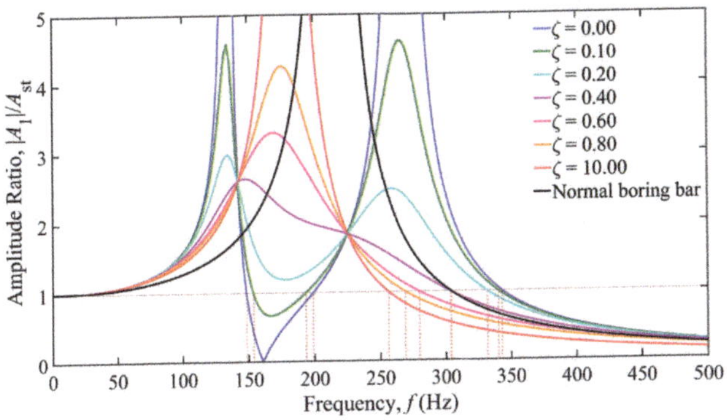

Figure 7. Amplitude ratio of the boring bar under different damping of the TDVA.

Based on the frequency-damping coupled three-dimensional response surface shown in Figure 8a, three characteristic vibration zones of the boring bar can be clearly identified: Area A contains three saddle-shaped peaks corresponding to the resonance bands of the first-order mode 85 Hz, second-order mode 165 Hz of the boring bar-tool system, and the TDVA-coupled mode 285 Hz, with amplitude ratios exceeding a value of five at these peaks. Adjustable Area C exhibits a saddle-valley structure, where the amplitude ratio remains below a value of one for damping ratios ζ = 0–0.15, enabling stable machining within the 149–198 Hz frequency range 49 Hz bandwidth. Area C occupies the geometric region bounded by nodal points C, D, E and the characteristic red response curve, as illustrated in Figure 8b. Area B is dominated by energy dissipation, where the amplitude ratio decays gradiently with coordinated increases in frequency and damping ratio; beyond 345 Hz, the amplitude ratio remains below a value of one across the entire range without requiring adjustments. Area B is confined to the right-lateral domain demarcated by reference points F, G, and the red line, as depicted in Figure 8b. Notably, in the 265–343 Hz transition band, increasing ζ to 0.4–0.6 reduces the amplitude ratio by over 50%, validating the damping ratio's optimization capability for edge frequency bands. Once the frequency exceeds 345 Hz, the system autonomously enters an inertial stability state regardless of ζ. This three-dimensional characteristic demonstrates that TDVA damping adjustments dynamically tailor energy dissipation pathways across frequency domains, achieving comprehensive stability enhancement in machining processes.

Based on the frequency–band vibration response analysis in Figure 9, the adaptation strategies for TDVA damping parameters exhibit significant variability. In the low-frequency band of 10–140 Hz, Figure 9a, the amplitude ratio is highly insensitive to damping ratio variations and consistently exceeds the chatter threshold, necessitating combined stiffness adjustments or process parameter optimization to avoid vibrations. When the excitation frequency increases to 141–198 Hz Figure 9b,c, the system enters a damping-sensitive region. Low damping ratios ζ reduce the amplitude ratio below 0.8, forming a stable machining window of a 49 Hz bandwidth. Notably, within the narrow 141–149 Hz sub-band, the amplitude ratio's sensitivity to ζ increases abruptly—adjusting ζ from 0.1 to 0.15 alone reduces the amplitude ratio by 300%. In the high-frequency band 265–343 Hz, increasing ζ is required to suppress secondary resonance. Once the frequency exceeds 345 Hz, the system enters an inertial stability domain, where the amplitude ratio autonomously converges to 0.4–0.5, independent of damping.

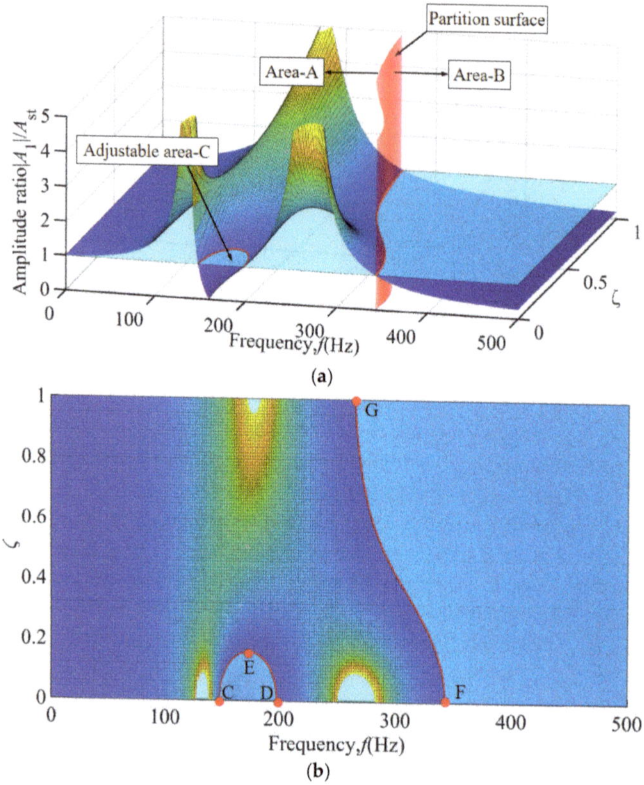

Figure 8. The amplitude ratio of the three-dimensional surface under the coupling effect of frequency and TDVA damping. (**a**) Macroscopic 3D surface diagram; (**b**) Top view of 3D curved surface.

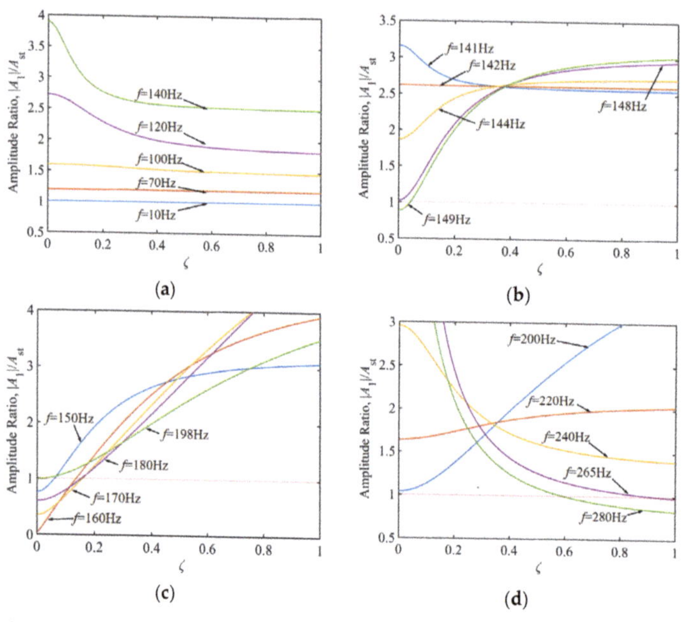

Figure 9. *Cont.*

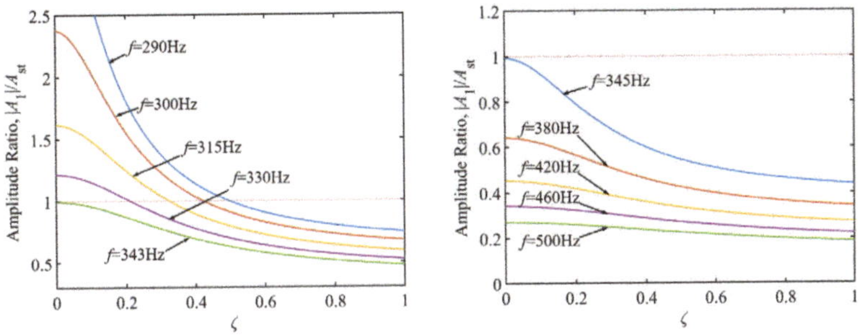

(e) (f)

Figure 9. Control characteristics of TDVA damping in frequency division on the amplitude ratio of boring bar. (**a**) Excitation frequency 10–140 Hz; (**b**) Excitation frequency 141–149 Hz; (**c**) Excitation frequency 150–198 Hz;(**d**) Excitation frequency 200–280 Hz; (**e**) Excitation frequency 290–343 Hz; (**f**) Excitation frequency 345–500 Hz.

4. Analysis of Boring Bar Vibration Characteristics Under Combined Stiffness–Damping Effects of TDVA

4.1. Stiffness Simulation Experiment of TDVA

Based on the stiffness regulation mechanism of the TDVA illustrated in Figure 3, the radial stiffness of the rubber bushing is dynamically adjusted via axial compression. To establish the quantitative relationship between stiffness and axial compression, a finite element model (FEM) of the rubber bushing was developed for static simulation. Under fixed constraints on the mass block, the equivalent stiffness was calculated using Hooke's law by controlling the displacement of the axial compression block and measuring the radial deformation Δx. The objective was to determine the relationship between radial stiffness and axial compression distance for the TDVA [31].

A simplified finite element model of a single rubber bushing (φ30 × 6 mm, outer diameter × wire diameter) was developed in ANSYS Workbench 2021 R2, as illustrated in Figure 10. The rubber material was modeled using the Mooney–Rivlin hyperelastic constitutive law to characterize its nonlinear mechanical behavior, whereas the mass block, axial compression block, and boring bar inner wall were defined as rigid components. Sliding contact interfaces were established between the rubber bushing and the rigid elements (mass block, axial compression block, and boring bar). Boundary conditions included full fixation of the mass block, rightward displacement loading on the axial compression block, and the subsequent application of a 200 N radial force to the rubber bushing. Owing to structural symmetry, only the upper half of the assembly was analyzed computationally. A refined hexahedral mesh with an element size of 0.05 mm was applied exclusively to the rubber bushing, while rigid components were exempted from meshing. Axial compression displacements were incrementally imposed to simulate varying operational conditions, with radial displacement (Δx) monitored synchronously. The equivalent stiffness of the rubber bushing was calculated via Hooke's law:

$$k = \frac{F}{\Delta x} \tag{7}$$

Since the TDVA incorporates two rubber bushings, its total stiffness is given by

$$k_2 = 2k \tag{8}$$

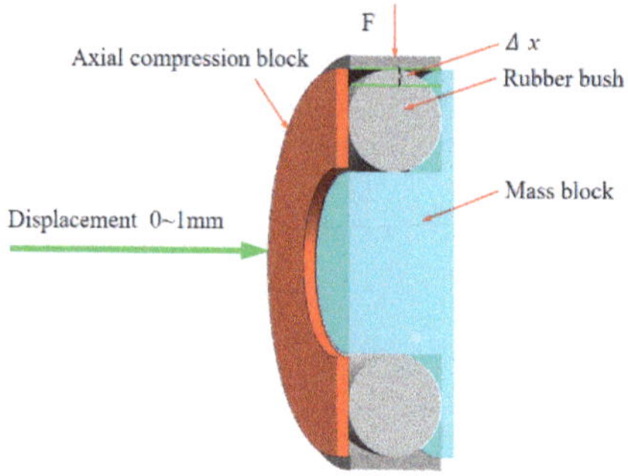

Figure 10. Stiffness analysis finite element model.

The simulation results are shown in Figure 11. Analysis of the simulated data reveals that the rubber material exhibits distinct stage-dependent mechanical behavior during axial compression. When the compression displacement is within 0.7 mm, the system stiffness follows a linear response regime, specifically a 12% increase in radial stiffness per 0.1 mm increment in compression. During this phase, the internal molecular chains of the material maintain a free conformational state, ensuring reversible energy absorption. As illustrated in Figure 12, which plots the relationship between TDVA stiffness and axial compression, surpassing the critical threshold of 0.7 mm axial compression triggers the rubber bushing's densification deformation phase. In this stage, the rate of change in Δx (radial displacement) decreases significantly. The directional rearrangement of molecular chains within the material induces a nonlinear stiffness surge. This intense deformation process not only accelerates fatigue damage accumulation but also causes irreversible plastic deformation. For engineering applications, it is recommended to strictly limit axial compression to within the 0.7 mm threshold to ensure structural stability and prolong component service life.

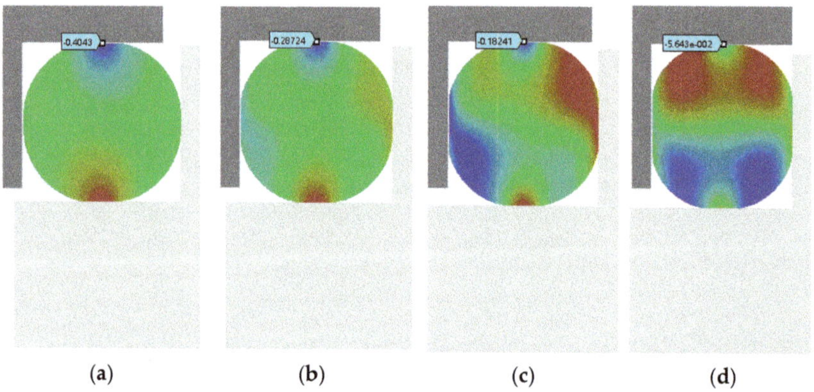

Figure 11. TDVA radial stiffness analysis. (**a**) Axial compression 0.1 mm; (**b**) Axial compression 0.4 mm; (**c**) Axial compression 0.6 mm; (**d**) Axial compression 1 mm.

This section establishes a constitutive relationship model between the radial stiffness k_2 of the TDVA and the axial compression distance. In engineering applications, the equivalent stiffness k_2 can be determined via swept-frequency excitation tests, and the optimal compression displacement can be back-calculated based on characteristic curves. When the axial compression is controlled within the 0–0.7 mm range, the stiffness adjustment range reaches 243%. Notably, the current stiffness regulation model does not account for time-varying damping effects. Under practical operating conditions, each 0.1 mm increase in axial compression alters the equivalent damping ratio, thereby affecting vibration control bandwidth. In the next phase, a coupled model of axial compression and damping is developed to quantitatively analyze the modulation mechanisms of compression displacement on damping characteristics.

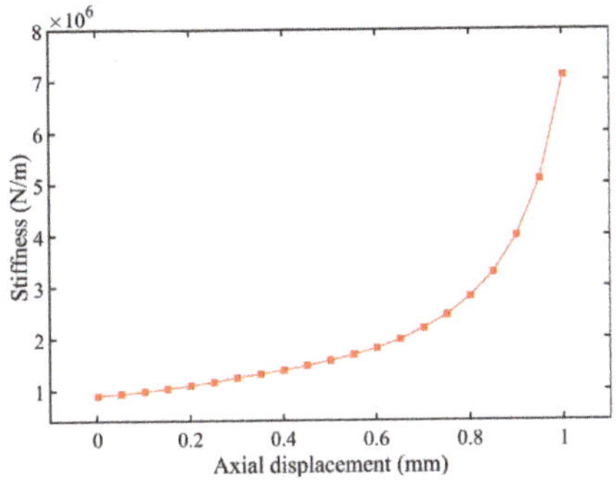

Figure 12. The relationship between stiffness and axial compression value of TDVA.

4.2. Damping Simulation Experiment of TDVA

The damping characteristics of the TDVA were quantitatively analyzed through coupled axial compression–radial loading simulations. Based on the finite element model (FEM) shown in Figure 13, periodic compressive displacements of 0.1 mm were applied radially under varying axial compression levels. The force–displacement curves of the rubber bushing formed hysteresis loops, where the enclosed area directly quantified the system's energy dissipation capacity. Simulation results indicate that as the axial compression displacement increased from 0 to 0.5 mm, the hysteresis loop area exhibited a nonlinear increasing trend. This trend reveals a directional modulation mechanism of axial compression on the TDVA's damping properties, providing a theoretical foundation for subsequent stiffness–damping coupling optimization.

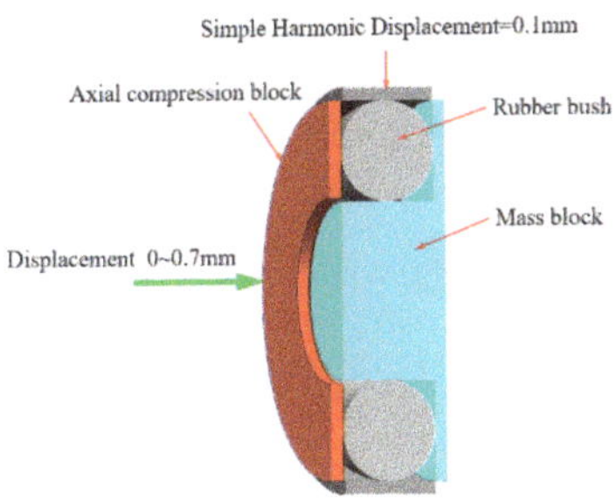

Figure 13. Damping analysis finite element model.

Equation (1) describes the dynamic model of the boring bar integrated with the Tunable Dynamic Vibration Absorber (TDVA). To investigate the damping characteristics of the TDVA, the system is modeled as a viscously damped oscillator subjected to external harmonic excitation. The equation of motion for a unit mass within this framework can be expressed as [32,33]

$$\ddot{x} + 2\zeta\omega_n\dot{x} + {\omega_n}^2 x = {\omega_n}^2 u(t) \tag{9}$$

where ω_n is the natural frequency of the TDVA and $u(t)$ is the sinusoidal excitation function. Within one cycle, the energy dissipation per unit mass ΔU (i.e., damping capacity) is defined as the energy consumed during a complete periodic motion:

$$\Delta U = 2\pi x_0^2 \omega_n \omega \zeta \tag{10}$$

where x_0 is the amplitude and ω is the excitation frequency. The maximum potential energy of the system per unit mass is

$$U_{\max} = \frac{1}{2}\frac{k}{m}x_0^2 = \frac{\omega_n^2 x_0^2}{2} \tag{11}$$

If the initial total energy of the system is denoted as $U_{\max}$, the loss factor η equals the specific damping capacity per radian over one damping cycle:

$$\eta = \frac{\Delta U}{2\pi U_{\max}} \tag{12}$$

From Equation (12), the loss factor for a simple harmonic oscillator with viscous damping is expressed as

$$\eta = \frac{2\pi x_0^2 \omega_n \omega \zeta}{2\pi \times \frac{\omega_n^2 x_0^2}{2}} = \frac{2\omega\zeta}{\omega_n} \tag{13}$$

For damped decay systems and forced vibrations, the most intense vibration response occurs when the excitation frequency approximates the natural frequency of the vibration absorber ($\omega \approx \omega_n$), where energy dissipation must be considered.

The above derivations assume mass normalization. Equation (10) represents the work performed by a unit mass to overcome resistance during one load–unload cycle, leading to

$$2\omega_n \zeta = c/m \tag{14}$$

where c is the viscous damping coefficient and mm is the mass. The unit mass and the energy dissipated by each hysteresis loop are equal to

$$\Delta U = 2\pi x_0^2 \omega c/m \tag{15}$$

For non-normalized systems, the energy dissipated per hysteresis loop by viscous damping is

$$\Delta U_v = 2\pi x_0^2 \omega c \tag{16}$$

The initial maximum energy can be expressed using the initial maximum potential energy:

$$U_{\max} = \frac{1}{2}kx_0^2 \tag{17}$$

The loss factor for non-normalized systems is derived as

$$\eta = \frac{\Delta U_v}{2\pi U_{\max}} = \frac{2\pi x_0^2 \omega c}{2\pi \times \frac{1}{2}kx_0^2} = \frac{2\omega c}{k} \tag{18}$$

From Equation (13), the damping ratio ζ is obtained:

$$\zeta = \frac{\omega c}{k} \tag{19}$$

According to the quantitative analysis of the hysteresis loop area in Figure 14, combined with Equation (16), the known amplitude x_0, and stiffness k, the parameter ωc can be solved, and the damping ratio ζ is subsequently derived using Equation (19). The damping ratio evolution curve in Figure 15 illustrates that within the axial compression range of 0–0.5 mm, ζ increases approximately linearly with compression displacement, exhibiting a 16% enhancement in energy dissipation efficiency per 0.1 mm increment. When the compression exceeds 0.5 mm, material nonlinearity reduces the slope of ζ growth by 25%, with an inflection point observed at 0.7 mm. Consequently, it is recommended to limit axial compression to the 0–0.5 mm range to achieve linear and controllable adjustment of ζ between 0.1 and 0.8. This design range ensures machining stability while avoiding stress relaxation failure of the rubber bushing under high compression.

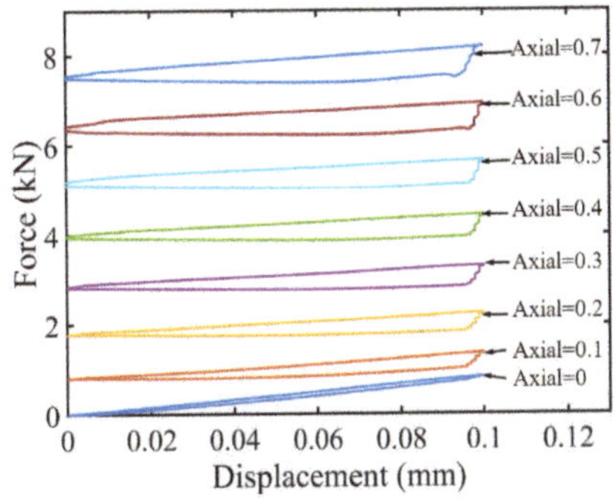

Figure 14. TDVA damping force and displacement hysteresis curve.

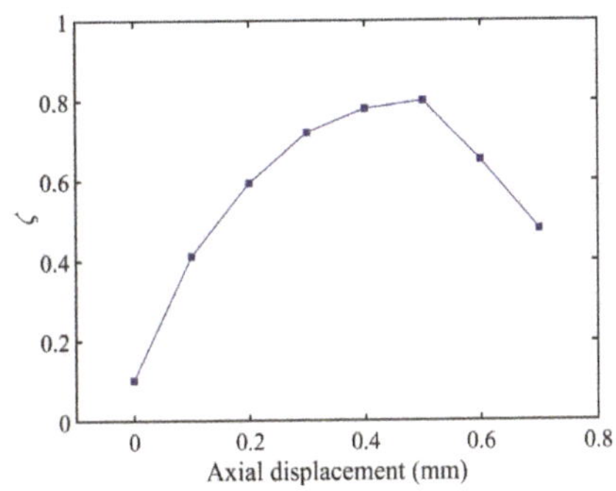

Figure 15. The relationship between damping and axial compression value of TDVA.

5. TDVA Stiffness and Damping Combined Effect

As established in prior sections, when the axial compression block in the TDVA moves rightward, compressing the rubber bushings, both the stiffness and damping of the TDVA change. By linking the results from Sections 4.1 and 4.2 under axial compression, the stiffness and damping exhibit approximately linear correlation within the 0.5 mm axial compression range, as illustrated in Figure 16.

In Figure 16a, the cyan plane represents an amplitude ratio of one. The intersection of the 3D surface with this plane is marked by red curves. Regions where the 3D surface lies below the amplitude ratio = 1 plane indicate favorable machining conditions with effective vibration suppression, termed Adjustable Area A and Adjustable Area B. Figure 16b and the top view of Figure 16a clearly show that in Machinable Zone A, spanning 163–215 Hz excitation frequency, vibration reduction is achievable with minimal axial compression. In Machinable Zone B (245–343 Hz), adjusting the axial compression of the TDVA enables the amplitude ratio to remain below a value of one, demonstrating the vibration-tuning capability of the proposed boring bar design. Beyond 343 Hz, the system's inherent stability autonomously converges the amplitude ratio to within a value of one, fulfilling machining requirements without parameter adjustments. This characteristic validates the TDVA boring bar's dynamic adaptability across the full frequency spectrum.

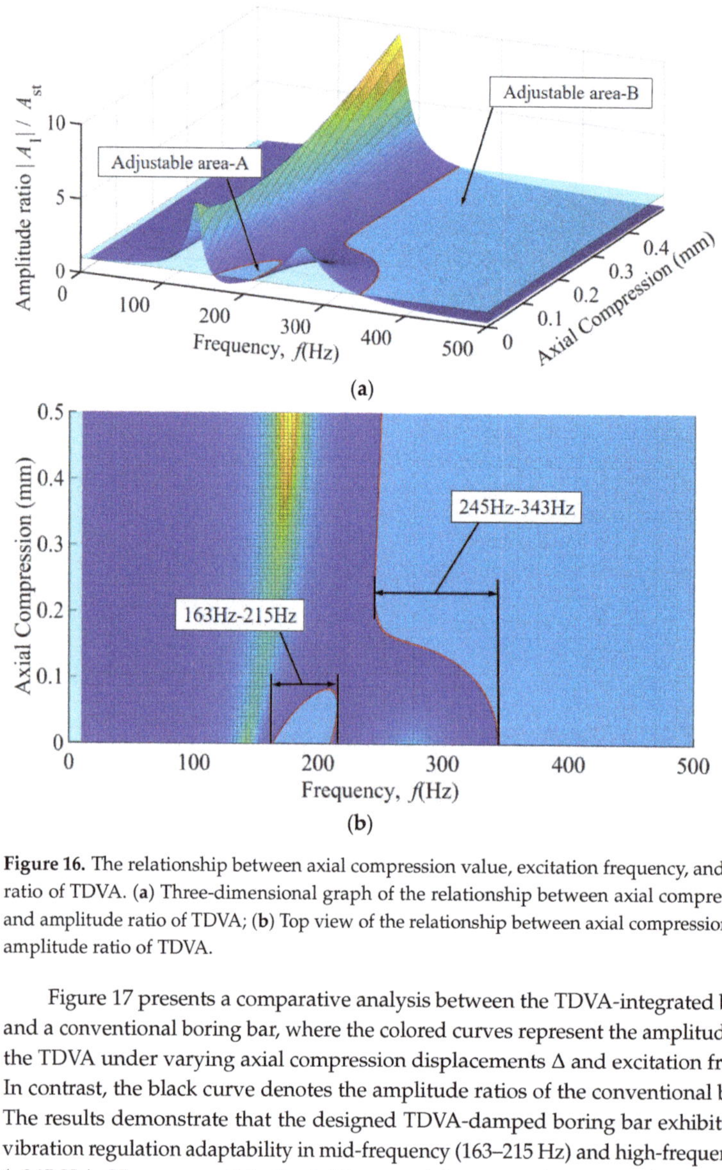

Figure 16. The relationship between axial compression value, excitation frequency, and amplitude ratio of TDVA. (**a**) Three-dimensional graph of the relationship between axial compression value and amplitude ratio of TDVA; (**b**) Top view of the relationship between axial compression value and amplitude ratio of TDVA.

Figure 17 presents a comparative analysis between the TDVA-integrated boring bar and a conventional boring bar, where the colored curves represent the amplitude ratios of the TDVA under varying axial compression displacements Δ and excitation frequencies. In contrast, the black curve denotes the amplitude ratios of the conventional boring bar. The results demonstrate that the designed TDVA-damped boring bar exhibits superior vibration regulation adaptability in mid-frequency (163–215 Hz) and high-frequency ranges (>245 Hz). However, within the 0–0.1 mm axial compression range, the curves display significant irregular fluctuations, necessitating supplemental Figure 18 to further clarify the tunability within the 163–215 Hz excitation band. For the conventional boring bar, the amplitude ratio (black curve) consistently exceeds a value of one at excitation frequencies below 304 Hz, indicating its operational incapacity and lack of tunability in low-frequency regimes. In contrast, the TDVA-damped boring bar achieves stable machining performance in both the 163–215 Hz and >245 Hz frequency ranges, effectively overcoming the limitations of traditional designs.

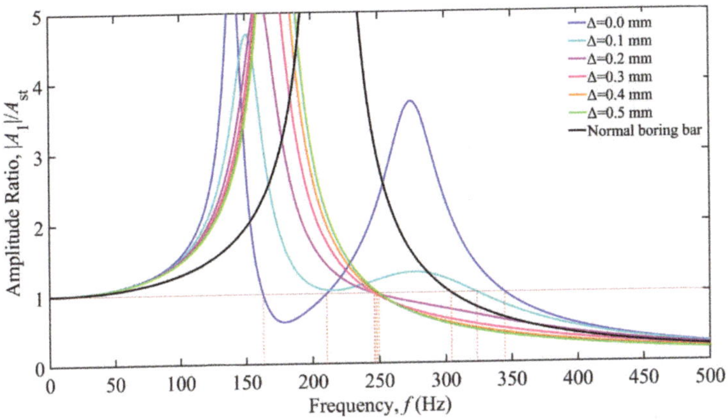

Figure 17. Comparison between TDVA boring bar and ordinary boring bar.

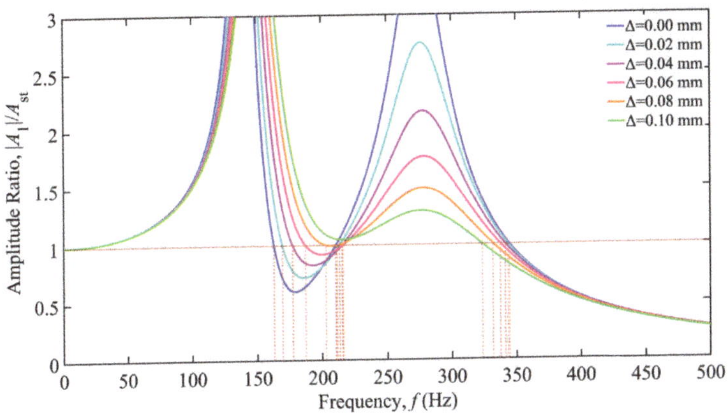

Figure 18. Time frequency domain focusing analysis of TDVA axial compression 0–0.1 mm.

6. Discussion

The proposed Tunable Dynamic Vibration Absorber (TDVA) boring bar overcomes the limitations of conventional passive DVA systems through a novel axial compression mechanism (0–0.5 mm range) that synergistically modulates both stiffness and damping parameters. Unlike traditional DVAs with fixed parameters optimized for narrow frequency bands, the TDVA enables real-time adaptation to variable machining conditions. Axial compression adjusts radial stiffness linearly at 5×10^4 N/m per 0.1 mm displacement while simultaneously tuning damping ratios from 0.1 to 0.8 in a quasi-linear regime (0–0.5 mm compression). Finite element simulations demonstrate that in this coupled regulation mechanism, stiffness increases by 12% per 0.1 mm compression, while damping evolves through distinct phases—controlled quasi-linear growth below 0.5 mm and nonlinear saturation beyond this threshold. These coordinated adjustments suppress vibrations by dynamically tailoring system impedance, achieving adjustable advantages compared to static DVA configurations.

The TDVA's dual-parameter tuning capability resolves two critical challenges in deep-hole boring: (1) mid-frequency vibration attenuation (163–215 Hz) via viscoelastic damping enhancement at low compression (0.1–0.3 mm) and (2) high-frequency stability (245–343 Hz) through stiffness-dominated resonance avoidance at higher compression (0.3–0.5 mm). This real-time adaptability extends the effective machining bandwidth by 28% compared to ordinary boring bars, outperforming passive DVAs and active solutions in reliability and simplicity. As shown in Table 2, the comparison of two types of boring bars in the 0–500 Hz range is presented. However, limitations persist: axial compression exceeding 0.5 mm induces nonlinear damping saturation and accelerated rubber bushing fatigue, restricting practical applications to the 0–0.5 mm range. Future research should prioritize fatigue-resistant composites and closed-loop control systems to exploit the full potential of this stiffness-damping synergy while addressing multi-axis vibration coupling.

Table 2. Comparison between TDVA boring bar and normal boring bar.

Boring Bar Type	Equivalent Stiffness (10^6 N/m)	Damping Ratio of the TDVA, ζ	Stiffness of the TDVA (106 N/m)	Processing Frequency Range (Hz)
TDVA boring bar	2.2057	0-1-0.8	9-16	163–215, 245–500
Normal boring bar	2.2134	-	-	304–500

7. Conclusions

This study reveals the stiffness–damping synergistic regulation mechanism by establishing a dynamic mapping relationship between the axial compression displacement of the rubber bushing and the vibration absorber parameters. Finite element simulations demonstrate that within the axial compression range of 0–0.7 mm, the radial stiffness increases linearly at a gradient of approximately 5×10^4 N/m. In comparison, the damping ratio exhibits nonlinear attenuation beyond 0.5 mm. The positive stiffness gradient regulation and negative damping gradient evolution form the theoretical foundation for decoupled dual-parameter control, offering an innovative solution for broadband vibration suppression through a single mechanical adjustment.

According to three-dimensional response surface analysis, the TDVA exhibits exceptional vibration suppression adaptability across the 163–343 Hz wide-frequency domain. For mid–low-frequency vibrations, 163–215 Hz, a micro-compression displacement of 0–0.1 mm combined with a stiffness of 7–7.5 $\times$ 10^5 N/m and a damping ratio of 0.1–0.4 stabilizes the amplitude ratio below a value of one. In the high-frequency range 245–343 Hz, a compression displacement of 0.2–0.5 mm with high stiffness 7.5–16 $\times$ 10^5 N/m and damping ratios 0.4–0.8 suppresses the amplitude ratio below a value of one, achieving over 65% reduction compared to conventional boring bars and overcoming their nonadjustable limitations below 303 Hz. With a reference excitation frequency range of 0–500 Hz, the TDVA boring bar achieves a 28% expansion in effective machining bandwidth.

The primary contribution of this work lies in the development of a mechanically tunable vibration absorber that simultaneously adjusts stiffness and damping via a single axial compression input, overcoming the limitations of conventional passive DVAs with fixed parameters. Unlike existing active solutions requiring complex control systems, the TDVA achieves real-time adaptability through purely mechanical modulation, ensuring robustness in industrial environments.

Author Contributions: Conceptualization, Y.G. and G.Y.; methodology, Y.G., G.Y. and D.X.; software, Y.G., Q.H. and J.X.; validation, Y.G., Q.H. and P.L.; formal analysis, Y.G.; investigation, G.Y.; writing—original draft preparation, Y.G.; writing—review and editing, Y.G. and J.X.; funding acquisition, Y.G. All authors have read and agreed to the published version of the manuscript.

Funding: This work was funded by the Scientific and technological research project of Heilongjiang Province (Grant No.2022ZXJ01A02, 2022ZXJ01A01); the Key Research and Development Program of Heilongjiang Province (Grant No. 2024ZXDXA36) and the Basic scientific research operating expenses project of Heilongjiang Province (Grant No. 145209405).

Institutional Review Board Statement: Not applicable.

Informed Consent Statement: Not applicable.

Data Availability Statement: The original contributions presented in this study are included in the article. Further inquiries can be directed to the corresponding author.

Conflicts of Interest: Author Jiao Xu was employed by the company HE Harbin Power Plant Valve Company Limited. The remaining authors declare that the research was conducted in the absence of any commercial or financial relationships that could be construed as a potential conflict of interest.

References

1. Sun, J.; Sun, C.; Yan, Z. State-of-art, challenges, and outlook on deep hole boring: Chatter suppression, tool wear monitoring, and error measurement. *Int. J. Adv. Manuf. Technol.* **2025**, *136*, 2075–2105. [CrossRef]
2. Li, H.; Xia, Y.; Su, B. Ultrasonic vibration-assisted chatter suppression for deep hole boring of stainless steel. *Int. J. Adv. Manuf. Technol.* **2024**, *131*, 1691–1703. [CrossRef]
3. Suyama, D.; Diniz, A.; Pederiva, R. The use of carbide and particle-damped bars to increase tool overhang in the internal turning of hardened steel. *Int. J. Adv. Manuf. Technol.* **2016**, *86*, 2083–2092. [CrossRef]
4. Matsubara, A.; Maeda, M.; Yamaji, I. Vibration suppression of boring bar by piezoelectric actuators and LR circuit. *CIRP Ann.—Manuf. Technol.* **2014**, *63*, 373–376. [CrossRef]
5. Sajedi Pour, D.; Behbahani, S. Semi-active fuzzy control of machine tool chatter vibration using smart MR dampers. *Int. J. Adv. Manuf. Technol.* **2016**, *83*, 421–428. [CrossRef]
6. Ghorbani, S.; Rogov, V.; Carluccio, A. The effect of composite boring bars on vibration in machining process. *Int. J. Adv. Manuf. Technol.* **2019**, *105*, 1157–1174. [CrossRef]
7. Liu, X.; Liu, Q.; Wu, S. Analysis of the vibration characteristics and adjustment method of boring bar with a variable stiffness vibration absorber. *Int. J. Adv. Manuf. Technol.* **2017**, *98*, 95–105. [CrossRef]
8. Patel, A.; Yadav, A.; Law, M. Damped Chatter Resistant Boring Bar Integrated with an Absorber Working in Conjunction with an Eddy Current Damper. *J. Vib. Eng. Technol.* **2022**, *11*, 2013–2024. [CrossRef]
9. Gourc, E.; Seguy, S.; Michon, G. Quenching chatter instability in turning process with a vibro-impact nonlinear energy sink. *J. Sound Vib.* **2015**, *355*, 392–406. [CrossRef]
10. Li, L.; Sun, B.; Hua, H. Analysis of the Vibration Characteristics of a Boring Bar with a Variable Stiffness Dynamic Vibration Absorber. *Shock Vib.* **2019**, 1–13. [CrossRef]
11. Li, L.; Sun, B.; Hua, H. Nonlinear system modeling and damping implementation of a boring bar. *Int. J. Adv. Manuf. Technol.* **2019**, *104*, 921–930. [CrossRef]
12. Li, L.; Yang, D.; Cui, Y. Optimization of machining performance in deep hole boring: A study on cutting tool vibration and dynamic vibration absorber design. *Adv. Prod. Eng. Manag.* **2023**, *18*, 371–380. [CrossRef]
13. Iklodi, Z.; Barton, D.; Dombovari, Z. Bi-stability induced by motion limiting constraints on boring bar tuned mass dampers. *J. Sound Vib.* **2022**, *517*, 116538. [CrossRef]
14. Houck, L.; Schmitz, T.; Smith, K. A tuned holder for increased boring bar dynamic stiffness. *J. Manuf. Process.* **2011**, *13*, 24–29. [CrossRef]
15. Shi, H.; Ma, C.; Li, X.; Wang, G. Equivalent linearization of hyperelastic rubber rings in dynamic vibration absorber boring bar and simulation of its dynamic properties. *Eng. Res. Express* **2024**, *6*, 015517. [CrossRef]
16. Rubio, L.; Loya, J.; Miguélez, M. Optimization of passive vibration absorbers to reduce chatter in boring. *Mech. Syst. Signal Process.* **2013**, *41*, 691–704. [CrossRef]
17. Miguélez, M.; Rubio, L.; Loya, J. Improvement of chatter stability in boring operations with passive vibration absorbers. *Int. J. Mech. Sci.* **2010**, *52*, 1376–1384. [CrossRef]
18. Liu, Q.; Li., J.; Ma, J. Structure design and vibration control of vibration reduction boring bar with variable damping. *Int. J. Adv. Manuf. Technol.* **2024**, *134*, 5301–5319. [CrossRef]
19. Gokulu, T.; Defant, F.; Albertelli, P. Stability analysis of multi-insert rotating boring bar with stiffness variation. *J. Sound Vib.* **2024**, *586*, 118497. [CrossRef]
20. Tian, J.; Wu, D.; Li, J. Design and Energy Dissipation Analysis of a Boring Bar Based on Particle Damping. *J. Vib. Eng. Technol.* **2025**, *13*, 51. [CrossRef]

21. Guo, X.; Zhu, Y.; Luo, Z. Variable stiffness tuned particle dampers for vibration control of cantilever boring bars. *Appl. Math. Mech.* **2023**, *44*, 2163–2186. [CrossRef]
22. Ramu, G.; Sundaramoorthy, K.; Singaravelu, C. A novel hybrid particle damping in boring bar for effective machining and chatter suppression. *Sādhanā* **2024**, *49*, 276. [CrossRef]
23. Lawrance, G.; Paul, P.; Shylu, D. Effect of metallic substrate and rubber elastic materials over passive constrained layer damping on tool vibration during boring process. *J. Low Freq. Noise Vib. Act. Control* **2024**, *43*, 1139–1157. [CrossRef]
24. Biju, C.; Shunmugam, M. Development of a boring bar with magneto rheological fluid damping and assessment of its dynamic characteristics. *J. Vib. Control JVC* **2018**, *24*, 3094–3106. [CrossRef]
25. Chen, F.; Hanifzadegan, M.; Altintas, Y. Active Damping of Boring Bar Vibration with a Magnetic Actuator. *IEEE/ASME Trans. Mechatron.* **2015**, *20*, 2783–2794. [CrossRef]
26. Yamada, K.; Matsuhisa, H.; Utsuno, H. Enhancement of efficiency of vibration suppression using piezoelectric elements and LR circuit by amplification of electrical resonance. *J. Sound Vib.* **2014**, *333*, 1281–1301. [CrossRef]
27. Wang, X.; Yang, B.; Yu, H. Optimal Design and Experimental Study of a Multidynamic Vibration Absorber for Multifrequency Excitation. *J. Vib. Acoust.* **2017**, *139*, 031011. [CrossRef]
28. Yang, Y.; Muñoa, J.; Altintas, Y. Optimization of multiple tuned mass dampers to suppress machine tool chatter. *Int. J. Mach. Tools Manuf.* **2010**, *50*, 834–842. [CrossRef]
29. Febbo, M.; Vera, S. Optimization of a Two Degree of Freedom System Acting as a Dynamic Vibration Absorber. *J. Vib. Acoust.* **2008**, *130*, 011013. [CrossRef]
30. Zhang, Q.; Zhao, W.; Li, Y. Transverse vibration of the boring bar for BTA deep hole machining under stochastic excitation. *J. Mech. Sci. Technol.* **2023**, *37*, 5635–5648. [CrossRef]
31. Kępczak, N.; Bechciński, G.; Rosik, R. Operating Properties of Deep Hole Boring Tools with Modified Design. *Materials* **2024**, *17*, 1551. [CrossRef] [PubMed]
32. Saffury, J.; Altus, E. Optimized chatter resistance of viscoelastic turning bars. *J. Sound Vib.* **2009**, *324*, 26–39. [CrossRef]
33. Zhang, Y.; Ren, Y.; Tian, J. Chatter stability of the constrained layer damping composite boring bar in cutting process. *J. Vib. Control* **2019**, *25*, 2204–2214.

materials

MDPI

Article

A Study on the Failure Behavior and Force Transmission of Composite Skin-Stringer Structures Under a Compressive Load

Guoyang Zhao [1,†], Jian Shi [2,*,†], Wei Xu [1], Nan Sun [3], Jianjiang Zeng [4,*], Guang Yang [5], Kun Song [6] and Jie Zheng [7,*]

1 School of Aviation Maintenance Engineering, Chengdu Aeronautic Polytechnic, Chengdu 610100, China; zhaoguoyang@cap.edu.cn (G.Z.); xuwei1395201@126.com (W.X.)
2 College of Aviation Engineering, Civil Aviation Flight University of China, Guanghan 618307, China
3 Department of Electrical and Computer Engineering, University of Alberta, Edmonton, AB T6G 2V4, Canada; nsun6@ualberta.ca
4 College of Aerospace Engineering, Nanjing University of Aeronautics and Astronautics, Nanjing 210016, China
5 School of Fiber Engineering and Equipment Technology, Jiangnan University, Wuxi 214000, China; gyang@nuaa.edu.cn
6 School of Mechanical and Power Engineering, Nanjing Tech University, Nanjing 211816, China; ksong@njtech.edu.cn
7 Department of Mechanical Engineering, University of Alberta, Edmonton, AB T6G 2R3, Canada
* Correspondence: shijian@cafuc.edu.cn (J.S.); zengjj@nuaa.edu.cn (J.Z.); jzheng11@ualberta.ca (J.Z.)
† These authors contributed equally to this work.

Abstract: Carbon fiber-reinforced composite stringers, which support aircraft skins in resisting tensile, compressive, and shear loads, are widely used in aircraft structures. These composite structures play a crucial role in enhancing the performance and safety of the structural integration of aircrafts. To better understand the load-bearing capacity of composite stringer structures, this study developed a novel model to study the complex failure and load transmission behavior of T800/3900S-2B fiber-reinforced composite skin-stringer structures under compressive loading. Compression strength tests were conducted on a composite stringer/skin structure, and a three-dimensional FEM was developed using Abaqus/Standard 2022. The model incorporated the modified 3D Hashin initiation criteria and Tserpes degradation law through a UMAT subroutine, which can effectively capture the in-plane ply failure and interlaminar damage. The results revealed a high degree of similarity between the load–displacement curves and failure modes (i.e., matrix compressive cracking, fiber compressive failure, and fiber–matrix shear-out failure) obtained from the simulations and those from the experiments. This study provides an efficient and accurate model to simulate the failure and load transfer of composite skin-stringer structures, offering significant advancements in understanding and predicting the behavior of these critical components.

Keywords: failure analysis; composite skin-stringer structures; load transfer; Hashin initiation criteria; Tserpes degradation law

1. Introduction

Carbon fiber-reinforced composite stringers are typically crucial longitudinal components within the wing structure of modern civil aircrafts. Functionally, they generally provide essential support to the aircraft skin, effectively preventing significant local deformation under external loads. Moreover, these stringers usually play a vital role in

https://doi.org/10.3390/ma18061380

enhancing the shear resistance capacity and compression stability of the skin. The composite H-shaped short stringer is often used as the main load-bearing structure. When subjected to compressive loads and subsequent damage, it frequently displays intricate alterations in force transmission paths and damage behavior, as reported in references [1–3]. Given these complex characteristics, the in-depth study of H-shaped short stringers is significant in aircraft structural design, as emphasized in references [4,5].

Pevzner et al. [6] improved an extended effective width method that is suitable for analyzing laminated composite stringer-stiffened circular cylindrical panels. Their study developed the relevant MATLAB code to analyze the bending buckling of the stiffeners, torsional buckling, combined bending and torsion buckling, and local buckling of the stringers. Moreover, the efficiency of this engineering method in the initial design stage was verified through comparison with experiments. Christian [7] introduced some engineering analysis methods regarding the buckling and post-buckling of thin-walled composite laminated beams. Badalló et al. [8] studied the optimization of a T-shaped stringer that is commonly used in CFRP-stiffened panels by employing three common genetic algorithms. The objectives of the optimization were to minimize the mass and maximize the critical buckling load, and the most suitable algorithm for that kind of problem was obtained. Wagner and Balzani [9] utilized the finite element method to investigate delamination and skin-stringer separation phenomena in stringer-stiffened fiber-reinforced composite shells. Vescovini et al. [10] analyzed the post-buckling response and failure of multi-stringer panels using the finite element method with three levels of approximation and obtained a deeper understanding of the effect of modeling parameters such as geometric and material parameters. Shi et al. [11–13] employed a semi-analytical multiscale algorithm to conduct progressive damage analysis on the composite stringers and the open-hole plates extracted from the webs of the stringers. This semi-analytical multiscale algorithm significantly enhanced the efficiency of multiscale calculations while ensuring computational accuracy. The analysis results obtained were in excellent agreement with the experimental results. Bnisch et al. [14] proposed a semi-analytical analysis method for the stress fields in composite skin-stringer junctions under a bending load. Compared with the results from finite element modeling, the stress results obtained by this semi-analytical approach were more accurate, and the computational time was only a small fraction of that required by finite element methods. Some fully analytical approaches [15–17] have been devised to analyze the post-buckling behavior of stiffened composite panels, which can rapidly calculate the failure loads. However, their principal shortcoming is that the accuracy of the result is compromised. The progressive damage method was employed in the literature [18,19] to conduct a detailed examination of the failure mechanisms and ultimate strength characteristics of composite stringers, and the analysis results exhibited a high degree of consistency with the experimental data. Recently, an increasing number of numerical studies [20–22] have been conducted to investigate the skin-stringer separation in stiffened composite panels and explore the influence of intra-laminar damages on the debonding evolution.

In summary, many researchers have studied the mechanical properties of either stringers or panels formed by stringers and skins. However, there are limited studies on the crushing behavior and force transmission paths of short co-cured skin-stringer structures under compressive loading. This paper performs experimental and numerical analyses of short H-shaped composite skin-stringer structures. The progressive damage method based on a modified 3D Hashin initiation criterion and Tserpes degradation law through a UMAT subroutine is used to predict the load–displacement curve, ultimate load, failure modes, and force transmission of the fiber-reinforced composite structures under compressive loading. Then, the analysis results are compared with the test results.

2. Failure Criteria and Material Property Degradation Rules

Composite structures exhibit complex and diverse failure modes. Failure criteria must account for stress coupling across material directions and differentiate damage states. Based on this aim, several failure criteria with excellent performance, such as the Chang-Chang [23,24], Hashin [25], and Puck [26] failure criteria, are widely acknowledged as notable and effective criteria. The macroscopic 3D Hashin failure criteria considering delamination and fiber–matrix shear-out [27] are used in the current study to determine the initial failure within the composite laminate of the stringer, as shown in Equations (1)–(7). It is generally believed that the mechanical behavior of the fiber is linearly elastic and brittle before fiber failure occurs. Therefore, the maximum stress criterion is adopted for the initial fiber failure.

The fiber tensile failure, for $\sigma_{11} > 0$, can be determined as follows:

$$\frac{\sigma_{11}}{X_T} \geq 1, \tag{1}$$

The fiber compressive failure, for $\sigma_{11} < 0$, can be determined as follows:

$$\frac{\sigma_{11}}{X_C} \geq 1, \tag{2}$$

The matrix tensile cracking, for $\sigma_{22} > 0$, can be determined as follows:

$$\left(\frac{\sigma_{22}}{Y_T}\right)^2 + \left(\frac{\sigma_{12}}{S_{12}}\right)^2 + \left(\frac{\sigma_{23}}{S_{23}}\right)^2 \geq 1, \tag{3}$$

The matrix compressive cracking, for $\sigma_{22} < 0$, can be determined as follows:

$$\left(\frac{\sigma_{22}}{Y_C}\right)^2 + \left(\frac{\sigma_{12}}{S_{12}}\right)^2 + \left(\frac{\sigma_{23}}{S_{23}}\right)^2 \geq 1, \tag{4}$$

The delamination in tension, for $\sigma_{33} > 0$, can be determined as follows:

$$\left(\frac{\sigma_{33}}{Z_T}\right)^2 + \left(\frac{\sigma_{13}}{S_{13}}\right)^2 + \left(\frac{\sigma_{23}}{S_{23}}\right)^2 \geq 1, \tag{5}$$

The delamination in compression, for $\sigma_{33} < 0$, can be determined as follows:

$$\left(\frac{\sigma_{33}}{Z_C}\right)^2 + \left(\frac{\sigma_{13}}{S_{13}}\right)^2 + \left(\frac{\sigma_{23}}{S_{23}}\right)^2 \geq 1, \tag{6}$$

The fiber–matrix shear-out, for $\sigma_{11} < 0$, can be determined as follows:

$$\left(\frac{\sigma_{11}}{X_C}\right)^2 + \left(\frac{\sigma_{12}}{S_{12}}\right)^2 + \left(\frac{\sigma_{13}}{S_{13}}\right)^2 \geq 1, \tag{7}$$

where σ_{ij} (i, j = 1, 2, 3) represents the normal and shear stress component; X_T, Y_T, and Z_T are the tensile strengths of the unidirectional ply in the normal direction. Similarly, X_C, Y_C, and Z_C denote the three normal compressive strengths. Additionally, S_{ij} (i, j = 1, 2, 3) is the shear strength of the unidirectional ply. Once the failure criteria are satisfied, the stiffnesses of the failed elements are reduced.

The sudden material property degradation rules of Tserpes [28,29] are adopted to reduce the material stiffness of the failed elements. Compared with other degradation schemes, these rules define a progressive three-dimensional stiffness degradation model based on the physical failure mechanism of the Chang-Chang model [23] and incorporate the mechanisms of delamination and fiber–matrix shear-out failure, as shown in Table 1. To avoid the singularity problem of the stiffness matrix during the calculation, 0 in the degradation criterion is set to a very small value of 10^{-9}.

Table 1. Sudden material property degradation rules.

Failure Mode	Sudden Degradation Rule
Fiber tensile failure	$E_{11}^{d} = 0.07 * E_{11}$
Fiber compressive failure	$E_{11}^{d} = 0.14 * E_{11}$
Matrix tensile cracking	$E_{22}^{d} = 0.2 * E_{22}, E_{12}^{d} = 0.2 * E_{12}, E_{23}^{d} = 0.2 * E_{23}$
Matrix compressive cracking	$E_{22}^{d} = 0.4 * E_{22}, E_{12}^{d} = 0.4 * E_{12}, E_{23}^{d} = 0.4 * E_{23}$
Fiber–matrix shear-out	$E_{12}^{d} = 0$
Delamination in tension and compression	$E_{33} = E_{23} = E_{13} = 0$

3. Uniaxial Compressive Experiment

The specimen of the stringer is composed of four parts: the top flange, web, bottom flange, and skin. The bottom flange and the skin are bonded together, and this bonded structure can be treated as an integrated laminated plate. The manufacturing processes and quality of these specimens are strictly controlled to reduce errors caused by possible defects, such as fiber discontinuities. To ensure a uniform load distribution, two resin-potted ends, each consisting of an aluminum alloy frame and a potting resin region, are attached at both ends of the specimen. The geometric dimensions of the specimen are presented in Figure 1. Specifically, the total length of the specimen measures 200 mm, which encompasses two 50 mm long reinforced sections and one 100 mm long test section. The cross-sectional widths of the top and bottom flanges are 24 mm and 50 mm, respectively, and the height of the web is 40 mm.

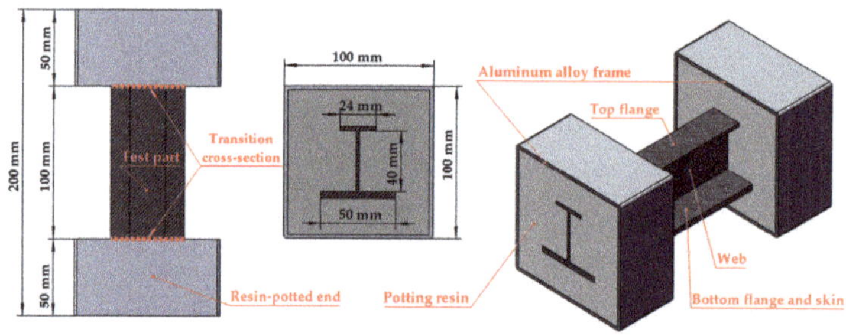

Figure 1. Geometric model of the test specimen, illustrating key structural components and dimensions. The specimen features a central test area with a transition cross-section, flanked by resin-potted ends for load application. A potting resin secures the structure, while an aluminum alloy frame provides external reinforcement. The internal H-section consists of a top flange, web, and bottom flange with skin.

Table 2 describes the stacking sequences, the thicknesses of the stringer components, and the number of plies. The thickness of each stringer ply is 0.19 mm. So, the thickness of each stringer component is equal to the number of plies multiplied by the thickness of a single ply.

Table 2. The section sizes and ply angles of the stringer.

Section	Ply Angles/°	Number of Plies	Thickness
Top flange	45/−45/0/−45/0/45/0/0/45/0/−45/0/−45/45	14	2.66 mm
Web	45/−45/0/−45/0/45/45/0/−45/0/−45/45	12	2.28 mm
Bottom flange	45/−45/0/45/0/−45/0/−45/45	9	1.71 mm
Skin	45/90/−45/0/90/45/−45/0/−45/45/90/0/−45/90/45	15	2.85 mm

The laminate material of the stringer is T800S/3900-2B, which is widely used in the main bearing structures of civil aircrafts. Table 3 presents the elastic modulus and strength parameters.

Table 3. Mechanical properties of T800S/3900-2B lamina, provided by Commercial Aircraft Corporation of China, Ltd., Shanghai, China.

Elastic Property	Value	Strength Property	Value
E_{11}	154 GPa	X_T	2690 MPa
E_{22}	9 GPa	X_C	1380 MPa
E_{33}	9 GPa	Y_T	93 MPa
G_{12}	7 GPa	Y_C	211 MPa
G_{23}	7 GPa	Z_T	93 MPa
G_{13}	7 GPa	Z_C	211 MPa
v_{12}, v_{23}, v_{13}	0.3	S_{12}, S_{23}, S_{13}	109 MPa

The uniaxial compression tests of the H-shaped stringer specimen were carried out on an Instron hydraulic test machine at a room temperature of 25 °C, and the displacement load was applied at a rate of 2 mm/min. Figure 2 presents the final failure state of the composite stringer, which showed that the catastrophic failures included delamination, fiber, and matrix compression failures. These failures occurred at the midpoint of the specimen, indicating that the resin-potted end design in this experiment effectively prevented a concentration of stress near the loading end caused by a non-uniform load application.

Figure 2. The final failure state of the stringer. The failure includes web buckling, flange delamination, and localized fractures, indicating progressive damage accumulation and ultimate load-bearing capacity loss.

Figure 3 shows the load–displacement curve of the uniaxial compression tests. During initial loading, some noises occurred at the resin-potted ends, which were likely generated as some small gaps within the potting region were being compacted. This corresponds to the stage of extremely low modulus at the initial part of the load–displacement curve. The specimens' average ultimate load is 166.15 kN, with a corresponding average displacement of 2.72 mm. Subsequently, a brittle fracture occurred in the middle of the specimen, resulting in a rapid load decline.

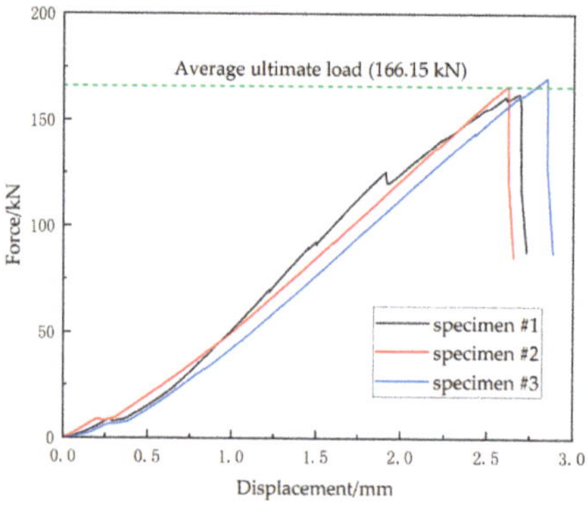

Figure 3. Load–displacement curves from the uniaxial compression test, showing the force–displacement response of three specimens. The applied forces and measured displacements are in the compressive direction. The average ultimate load is 166.15 kN, as marked by the dashed green line.

4. Finite Element Model Description

The element type C3D8R was selected for the FEM of the composite stringer. The element size of the two resin-potted ends was 1.5 mm, and each stringer ply was meshed separately. Specifically, the mesh sizes of each ply in the thickness direction and in-plane direction were set to 0.19 mm and 0.75 mm to balance computational efficiency and accuracy. The finite element model is presented in Figure 4. In the model, one end of the specimen was securely fixed, while a displacement-controlled load was applied to the other end. A reference point was set at the geometric center of the stringer's loading surface to ensure a uniform distribution of the load acting on the structure. Additionally, a multi-point constraint (MPC) was employed. This MPC tied the reference point to all the nodes of the loading surface. The stringer's axial displacement load was then applied to this reference point. This arrangement enabled the load to be applied and facilitated the monitoring and output of the load–displacement history data during the analysis process.

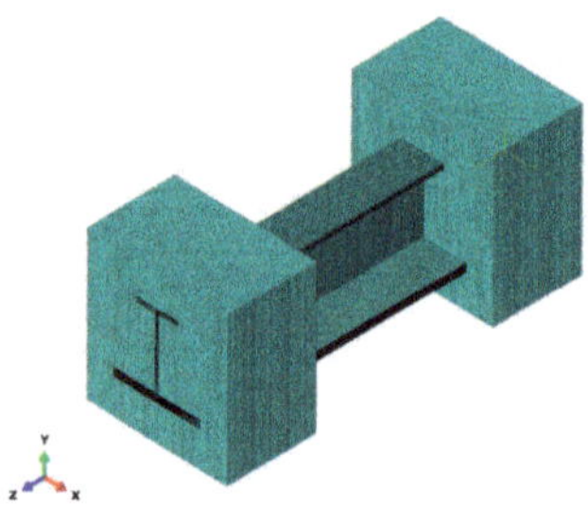

Figure 4. FEM mesh of stringer structure.

5. Results and Discussion

The three-dimensional FEM was established using Abaqus/Standard 2022, which integrated the modified 3D Hashin initiation criterion and Tserpes degradation law through a UMAT subroutine. Solution-dependent state variables (SDVs) within the Abaqus UMAT subroutine were used in the simulation. Specifically, the failure parameters are represented by these SDVs. SDV1 to SDV7 correspond to failure criterion Formulas (1)–(7). When an SDV is 0, its corresponding element is unfailed, and the value of 1 indicates that element failure has occurred.

Figure 5 shows the load–displacement curve derived from the simulation. The ultimate load of this curve is 168 kN. The results reveal a 1.11% relative error between the simulated and experimental average ultimate loads. Since the simulation model does not consider actual factors in the reinforced ends, such as the voids, the load–displacement curve exhibits more distinct linearity and increased brittleness during failure. The displacement corresponding to the ultimate load in the simulated curve is 2.42 mm, slightly smaller than its experimental counterpart, and this minor deviation is thus deemed reasonable.

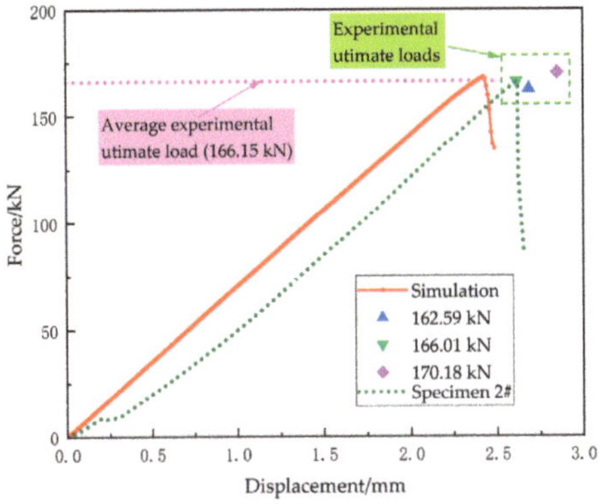

Figure 5. Load–displacement response from the uniaxial compression simulation. The simulated force–displacement curve (solid red line) is compared to the experimental results, including individual experimental ultimate loads (marked symbols) and the experimental load–displacement curve of specimen 2# (green dashed line). The average experimental ultimate load (166.15 kN) is indicated by the dashed purple line, demonstrating good consistency between the simulated and experimental data.

When the external load reaches 149 kN, initial compressive fiber damage occurs on the outer edges of the top and bottom flanges at the transition cross-sections of the stringer. The failure modes mainly include fiber compression failure in the 0° ply, matrix compression failure in the ±45° plies, and fiber–matrix shear-out failure in the 0° ply. The potting regions constrain all degrees of freedom (DOFs) of the two clamped parts of the stringer, except the longitudinal DOF. This constraint effectively avoids abnormal stress concentration at the contact regions between the stringer ends and the fixture block of the test machine. The maximum stress values occur at the transition cross-sections of the stringer. Among them, the longitudinal normal stress σ_{11} in the 0° ply is significantly larger than that in other plies. This indicates that the 0° plies mainly bear the compressive load. Therefore, initial fiber compression failure occurs first in the 0° plies, and the failure locations are at the transition cross-sections, as shown in Figure 6. Only a small number of elements in the top flange have failed, while more elements have failed in the 0° ply of the bottom flange and all 0° plies of the skin, and no fiber compression failure is observed in the web. The percentage of failed elements in the upper flange is nearly 20%, whereas failed elements of the bottom flange and skin contribute to approximately 80% of the total structural failure. The above-mentioned failure states indicate that the 0° plies of the bottom flange and the skin mainly bear the compressive load. Moreover, the bottom flange and the skin achieve the maximum thickness and the strongest load-bearing capacity, which proves that the force transmission path before the occurrence of initial damage is reasonable.

In the same simulation case, when the external load reaches 149 kN, some matrix compression failures are generated on both ends' cross-sections of the stringer test part, as shown in Figure 7. The initial matrix compression failure occurs earlier than the fiber compression failure, and there is a certain degree of damage propagation. The matrix compression damage mainly occurs in the non-0° plies of the skin and the bottom flange. The main reason is that these plies are subjected to a relatively large in-plane transverse compressive stress, σ_{22}, which is caused by the compressive load. A few elements have triggered matrix compression failures on the top flange, while no matrix failure is observed on the web. This situation also indicates that the bottom flange and the skin are the main parts bearing the compressive load.

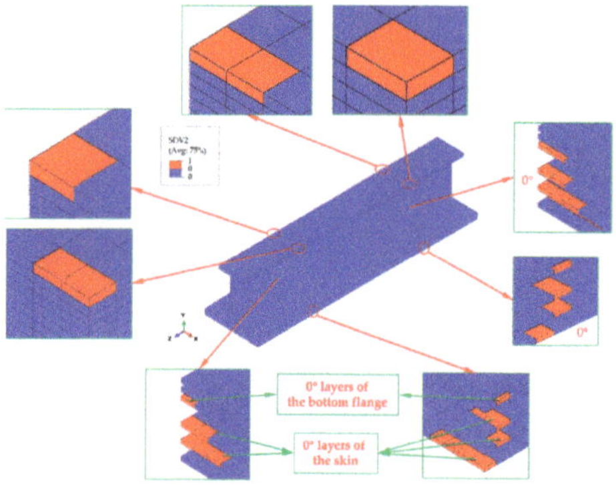

Figure 6. Initial fiber compression failure distribution in the stringer structure, highlighting affected 0° layers in the bottom flange and skin. The red regions indicate failed elements, showing failure initiation at the transition cross-sections. The insets provide a detailed view of localized damage, revealing stress concentration points leading to progressive failure.

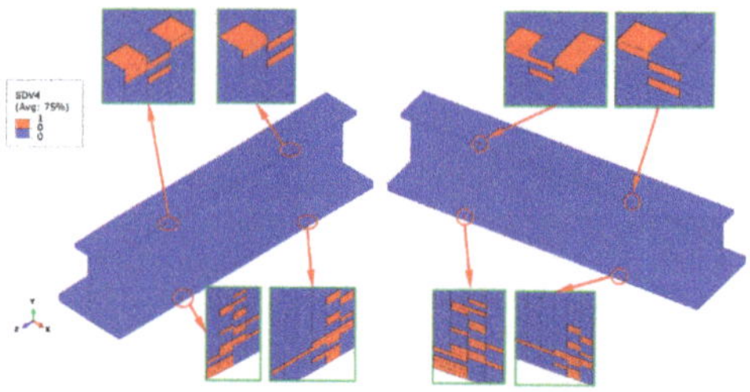

Figure 7. Matrix compressive failure distribution at a 149 kN load. The red regions indicate failed elements, located in the top flange, bottom flange, and skin near the transition areas. The insets provide a closer view of localized failures, showing progressive damage that is spread under increasing compressive loads.

In addition to spreading axially along the stringer, the matrix compression damage in the non-0° plies of the skin and the bottom flange also propagates transversely, as shown in Figure 8. The matrix failure initiates in the elements at the outer edges of the transition cross-sections, as shown in the section views A-A and B-B. The top and bottom flanges deform in thickness and the transverse directions under compression, with the web connecting them leading to a higher in-plane shear stress field. As the compressive load increases, the damage at the outer edges of the flange transition cross-sections gradually propagates towards the web.

The other failure modes mainly include the fiber–matrix shear-out failure in the initial failure stage, as shown in Figure 9. The shear failure elements are localized on the edges of the transition cross-sections of the stringer test part. The 0° ply of the top and bottom flanges experiences significant stress, σ_{11}, causing the stress components to satisfy the shear failure criteria and trigger fiber–matrix shear-out failure.

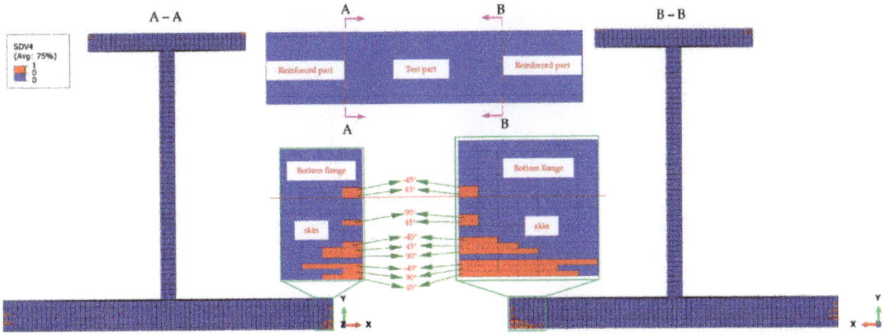

Figure 8. Matrix compressive failure distribution at the transition cross-section under a 149 kN load, highlighting damage progression in the bottom flange and skin. The red regions indicate failed elements, showing failure initiation at the outer edges and gradual propagation toward the web due to increased compressive stress. The insets provide a detailed view of localized damage.

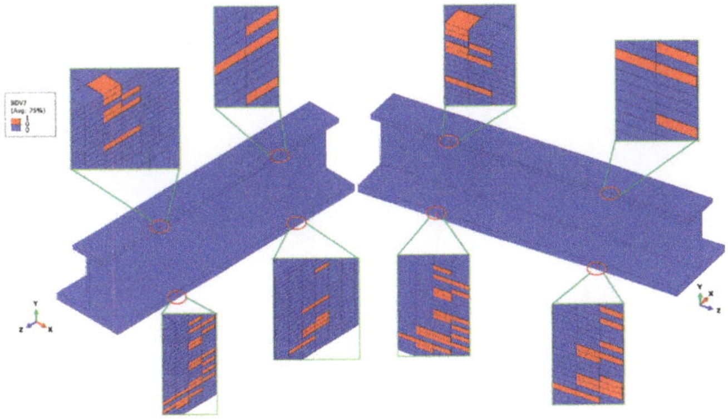

Figure 9. Fiber–matrix shear-out failure distribution at a 149 kN load, highlighting the failure concentration along the edges of the transition cross-sections. The red regions indicate failed elements. The insets provide a detailed view of the localized shear failure propagation, illustrating its role in structural degradation.

Under the ultimate load of 168 kN, the fiber compressive failure propagates extensively within the transition cross-sections at both ends of the stringer test part, as shown in the section views C-C of Figure 10. Fiber compressive failure occurs in most of the plies of the top flange cross-section, which causes the specimen to reach peak load. Relatively large areas with fiber failure also exist in the bottom flange and the skin, but the extent of the lateral damage is minimal, so they still retain a substantial load-bearing capacity. In the web of the stringer, however, there are only isolated regions with fiber compressive failure in the 0° plies. Therefore, the subsequent external load is mainly borne by the web, the bottom flange, and the skin.

At the ultimate load, the matrix compressive failure that mainly occurs in the ±45° plies also propagates extensively, with the most severe failure being observed at the top flange, as shown in the section view D-D of Figure 11. Extensive damage is also observed on both sides of the transition cross-sections of the stringer test part, but most of the plies have not been penetrated through the thickness. Owing to the extensive fiber compressive failure in the top flange, the load-bearing capacity of the top flange is significantly reduced, thus transferring the load to the web. Therefore, an elliptical area of matrix failure appears in the middle of the web, and the failure still mainly occurs in the ±45° plies, as shown in the section views E-E.

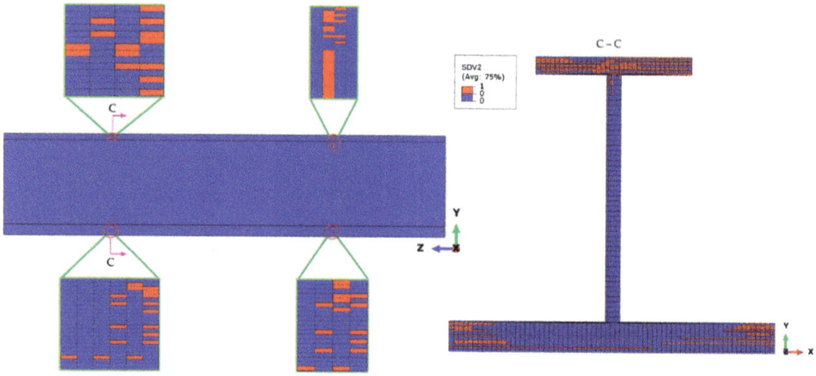

Figure 10. The fiber compressive failure distribution under the ultimate load of 168 kN, highlighting significant failure expansion in the transition cross-sections. The red regions indicate failed elements, primarily concentrated in the top flange, bottom flange, and skin, while the web experiences minimal failure. The insets provide a detailed view of the localized damage.

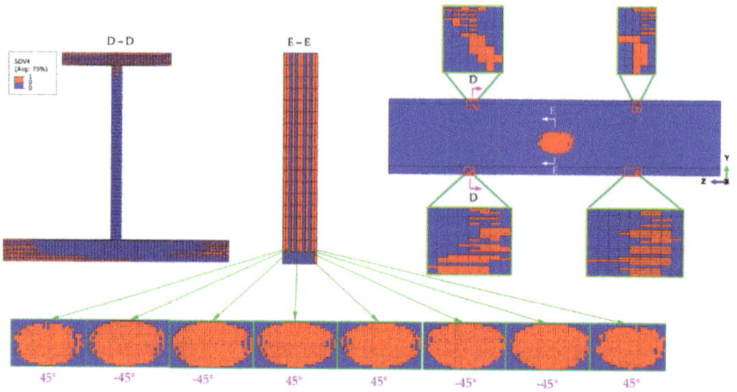

Figure 11. Matrix compressive failure distribution under the ultimate load, primarily affecting the ±45° plies. An elliptical failure zone forms in the middle of the web, as shown in the sectional views. The insets provide a detailed view of the localized damage, highlighting the failure propagation across the transition cross-sections.

Under the ultimate load, the fiber–matrix shear-out failure has propagated extensively throughout the transition cross-section of the stringer test part, as shown in the section views F-F of Figure 12. The most severe failure regions are located at the top flange, where the failure of most plies has penetrated the cross-section and extended to the upper end of the web. The fiber–matrix shear-out failure is comparable to the fiber compressive failure on the skin and the bottom flange in terms of the influence of the stress, σ_{11}, in the direction of the fiber to these failures. However, the fiber–matrix shear-out failure in this area has not penetrated the thickness yet.

After reaching the ultimate load, fiber compression failures, matrix compression failures, and fiber–matrix shear-out failures initiate at the root cross-section of the top flange test part, reducing the load-carrying capacity. Subsequently, the load-bearing function transfers rapidly to the web, bottom flange, and skin, causing damage in the FEM midsection to propagate rapidly and penetrate the entire model, forming the final failure state. Figure 13 shows the fiber compressive failure in the final state. Fiber compressive failures of the non-0° plies occur at the top and bottom root cross-sections of the stringer test part. A significant concentration of fiber compressive failures exists in the 0° ply at the middle of the specimen, as these layers primarily bear the compressive load. The location and form of the final fiber compressive failure are consistent with the results of the experiment.

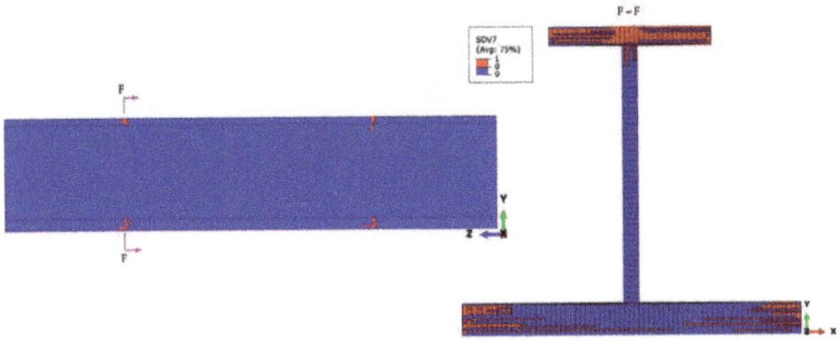

Figure 12. Fiber–matrix shear-out failure distribution under the ultimate load, highlighting the failure concentration at the transition cross-section, particularly in the top flange and bottom flange. The red regions indicate failed elements. Section F-F provides a detailed view, showing that while the failure has significantly progressed, it has not yet fully penetrated transversely.

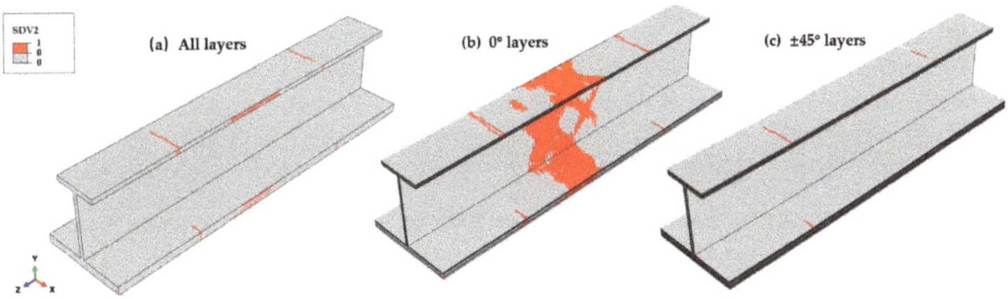

Figure 13. The final fiber compressive failure distribution across different ply orientations. (**a**) The failure distribution in all layers. (**b**) The 0° layers: extensive failure in the transition and central regions, where the compressive loads are primarily carried. (**c**) The ±45° layers: limited failure, mainly at transition points. The results indicate that 0° plies bear the majority of the compressive stress, leading to significant structural degradation.

Matrix tensile failures mainly occur in the 0° ply at the middle position of the stringer, as shown in Figure 14, reducing the elastic modulus, E_{11}, of the failed elements and causing a significant increase in strain in the fiber direction. Poisson's ratio effects elevate the transverse strain and stress in the failed elements, leading to extensive matrix tensile failures in the 0° ply at the specimen's midsection. The distribution of matrix tensile failures is in close agreement with that of fiber compressive failures. For the ±45° plies, matrix tensile failures are irregularly distributed on the web, with primary contributions arising from the shear stress components σ_{12} and σ_{23}.

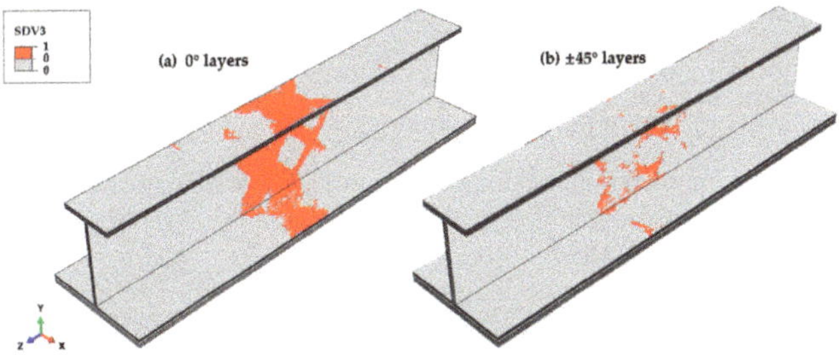

Figure 14. Final matrix tensile failure distribution across different ply orientations. (**a**) The 0° layers: extensive failure concentrated in the middle section. (**b**) The ±45° layers: failures are irregularly distributed across the web.

Matrix compressive failure rarely occurs in the 0° ply, yet it occurs extensively in the ±45° plies, as shown in Figure 15. In the ±45° plies, the matrix compressive failure initiates at the transition cross-section of the stringer test part. Subsequently, as the load approaches the ultimate load, isolated failure regions emerge in the web midsection. Shortly afterwards, these regions expand rapidly and penetrate the web thickness. Simultaneously, transverse penetration predominantly occurs in the ±45° plies of the top and bottom flanges, as well as the central region of the skin.

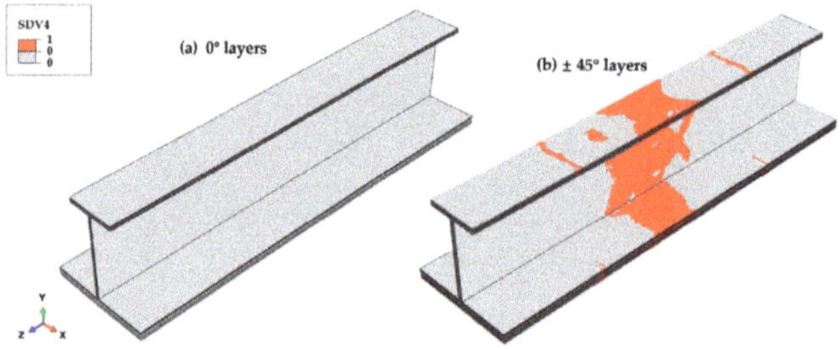

Figure 15. Final matrix compressive failure distribution across different ply orientations. (**a**) The 0° layers: minimal failure observed. (**b**) The ±45° layers: extensive failure concentrated in the transition cross-section and central region, while the damage propagates along the height of the web and transversely through the top and bottom flanges and skin in the central region of the structure.

The fiber–matrix shear-out failure is closely correlated with the stress along the fiber direction. As illustrated in Figure 16, the distribution pattern of the fiber–matrix shear-out failure in the 0° plies resembles that of the fiber compressive failure. While the longitudinal stress σ_{11} plays a role, the shear stresses σ_{12} and σ_{23} also drive the initiation of fiber–matrix shear-out failure. In the case of the ±45° plies, the fiber–matrix shear-out failure is localized at the transition cross-section of the test part and does not occur elsewhere. The analysis results indicate that delamination in tension and compression rarely occurs in the thickness direction due to the negligible normal stress σ_{33}. However, fiber–matrix shear-out failures predominate in the 0° plies, while the ±45° plies exhibit minimal damage. This indicates that the shear stress difference between the 0° and ±45° plies is the primary mechanism for shear delamination.

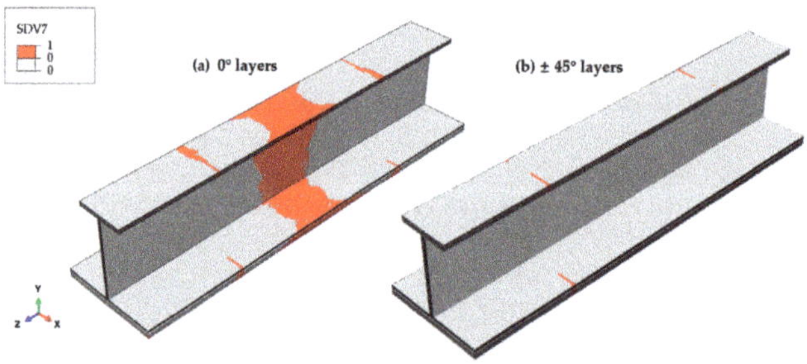

Figure 16. Final fiber–matrix shear-out failure distribution across different ply orientations. (**a**) The 0° layers: extensive failure in the transition and central regions. (**b**) The ±45° layers: failure is localized to the transition cross-section, without propagation to other regions.

6. Conclusions

Through carefully designed tests on the fiber-reinforced composite stringer and skin structure, the ultimate load, load–displacement curve, and failure modes were accurately determined. The load–displacement curve mainly exhibited brittle behavior, which is a crucial indicator of the material's mechanical response under compressive loading. The failure modes primarily included fiber compressive failure, matrix compressive failure, and fiber–matrix shear-out failure localized at the specimen's midsection, reflecting the complex interaction between the fiber and matrix components.

In addition to the experimental investigations, an in-depth simulation analysis was conducted to explore the damage evolution process further. Specifically, this analysis focused on three aspects: initial damage, damage under the ultimate load, and final failure state. Remarkably, the failure modes obtained from the simulation were in excellent agreement with the experimental results, validating the reliability of the simulation model. Specifically, the initial damage identified through simulation initiated at the transition cross-sections of the top and bottom flanges of the test section. As the external load gradually increased, extensive fiber and matrix compressive failures occurred at the transition cross-sections of the top flange. This accumulation of damage ultimately led to reaching the ultimate load. Consequently, isolated matrix compressive failure was observed at the web midsection. Finally, the damage propagated rapidly through the model midsection, closely matching the experimental final failure state.

The simulation process effectively revealed the intricate load transfer path during the damage progression. Due to the reinforcing effect of the potting-reinforced ends, the damage at the top flange transition cross-section of the test section was the most severe. After reaching the ultimate load, the load-bearing function was transferred to the specimen's midsection.

This study provides a solid foundation for improving potting-reinforced end designs and optimizing specimen configuration. For instance, a more sophisticated potting-reinforced end could be designed to effectively alleviate the concentration of stress. Moreover, the ply thickness or transverse dimension of the top flange could be appropriately adjusted to enhance its load-bearing capacity, thereby optimizing the overall performance of the composite structure.

Author Contributions: Conceptualization, G.Z., J.S., W.X., J.Z. (Jianjiang Zeng), G.Y. and J.Z. (Jie Zheng); Methodology, J.S., W.X., N.S., J.Z. (Jianjiang Zeng), G.Y., K.S. and J.Z. (Jie Zheng); Software, G.Z., J.S., W.X., G.Y., K.S. and J.Z. (Jie Zheng); Validation, G.Z., J.S., W.X., N.S., J.Z. (Jianjiang Zeng), K.S. and J.Z. (Jie Zheng); Formal analysis, G.Z., J.S., N.S., J.Z. (Jianjiang Zeng), K.S. and J.Z. (Jie Zheng); Investigation, G.Z., J.S., N.S. and J.Z. (Jie Zheng); Resources, J.S.; Data curation, G.Z., J.S., W.X., N.S., J.Z. (Jianjiang Zeng), G.Y., K.S. and J.Z. (Jie Zheng); Writing – original draft, G.Z. and J.S.; Writing—review & editing, J.Z. (Jie Zheng); Visualization, G.Z., J.S., W.X., N.S., G.Y. and J.Z. (Jie Zheng); Supervision, J.S., J.Z. (Jianjiang Zeng), G.Y., K.S. and J.Z. (Jie Zheng); Project administration, J.S.; Funding acquisition, J.S. All authors have read and agreed to the published version of the manuscript.

Funding: This research was funded by "the Fundamental Research Funds for the Central Universities", grant number PHD2023-013; "Research Center for Chengdu Aviation Industry Development and Cultural Construction", grant number CAIACDRCXM2024-03; and "Research on the characteristics of nitrogen-doped micro-arc oxidation coatings on titanium alloy surfaces", grant number ZZX0624088.

Institutional Review Board Statement: Not applicable.

Informed Consent Statement: Not applicable.

Data Availability Statement: The original contributions presented in this study are included in the article. Further inquiries can be directed to the corresponding authors.

Conflicts of Interest: The authors declare no conflicts of interest.

References

1. Liu, D.; Bai, R.; Wang, R.; Lei, Z.; Yan, C. Experimental study on compressive buckling behavior of J-stiffened composite panels. *Opt. Lasers Eng.* **2019**, *120*, 31–39.
2. Falkowicz, K.; Debski, H.; Wysmulski, P.; Rózyło, P. The behaviour of compressed plate with a central cut-out, made of composite in an asymmetrical arrangement of layers. *Compos. Struct.* **2019**, *214*, 406–413.
3. Sellitto, A.; Riccio, A.; Russo, A.; Zarrelli, M.; Toscano, C.; Lopresto, V. Compressive behaviour of a damaged omega stiffened panel: Damage detection and numerical analysis. *Compos. Struct.* **2019**, *209*, 300–316.
4. Xia, F.; Wang, Z.; Wang, Y.; Liu, H.; Xue, J. Post-Buckling Response of Carbon/Epoxy Laminates with Delamination under Quasi-Static Compression: Experiments and Numerical Simulations. *Materials* **2024**, *17*, 5047. [CrossRef]
5. Doan, Q.H.; Thai, D.K.; Tran, N.L. A Numerical Study of the Effect of Component Dimensions on the Critical Buckling Load of a GFRP Composite Strut under Uniaxial Compression. *Materials* **2020**, *13*, 931. [CrossRef]
6. Pevzner, P.; Abramovich, H.; Weller, T. Calculation of the collapse load of an axially compressed laminated composite stringer-stiffened curved panel—An engineering approach. *Compos. Struct.* **2008**, *83*, 341–353.
7. Christian, M. Buckling and post-buckling of thin-walled composite laminated beams—A review of engineering analysis methods. *Appl. Mech. Rev.* **2020**, *72*, 020802.
8. Badalló, P.; Trias, D.; Marín, L.; Mayugo, J.A. A comparative study of genetic algorithms for the multi-objective optimization of composite stringers under compression loads. *Compos. Part B Eng.* **2013**, *47*, 130–136.
9. Wagner, W.; Balzani, C. Simulation of delamination in stringer stiffened fiber-reinforced composite shells. *Comput. Struct.* **2008**, *86*, 930–939.
10. Vescovini, R.; Davila, C.G.; Bisagni, C. Failure analysis of composite multi-stringer panels using simplified models. *Compos. Part B* **2013**, *45*, 939–951.
11. Shi, J.; Tong, M.; Zhou, C.; Ye, C.; Wang, X. Progressive Failure Analysis in Open-Hole Tensile Composite Laminates of Airplane Stringers Based on Tests and Simulations. *Appl. Sci.* **2020**, *11*, 185. [CrossRef]
12. Shi, J.; Tong, M. Multi-Scale Nonlinear Progressive Damage and Failure Analysis for Open-Hole Composite Laminates. *Aerospace* **2022**, *9*, 59. [CrossRef]
13. Shi, J.; Zeng, J.; Zheng, J.; Shi, F.; Yang, G.; Tong, M. Multiscale Progressive Failure Analysis for Composite Stringers Subjected to Compressive Load. *Materials* **2024**, *17*, 3169. [CrossRef]
14. Bnisch, N.; Schilling, J.C.; Mittelstedt, C. Stress fields at skin-stringer junctions in composite aircraft fuselages. *Int. J. Mech. Sci.* **2024**, *284*, 109737. [CrossRef]
15. Mittelstedt, C.; Erdmann, H.; Schröder, K.U. Postbuckling of composite plates under inplane shear: Closed-form approximate solutions. *Arch. Appl. Mech.* **2011**, *81*, 1409–1426. [CrossRef]
16. Schilling, J.; Mittelstedt, C. Local postbuckling of omega-stringer-stiffened composite panels. *Thin-Walled Struct.* **2022**, *181*, 110027. [CrossRef]
17. Schilling, J.; Mittelstedt, C. Closed-form post-buckling analysis of shear deformable composite laminated panels. *Arch. Appl. Mech.* **2025**, *95*, 17. [CrossRef]
18. Chen, D.; Yan, R.; Xin, L.; Li, M. Progressive failure and ultimate strength analysis of sandwich composite hat-stringer-stiffened panels under transverse in-plane loading. *J. Mech. Mater. Struct.* **2021**, *16*, 1–21. [CrossRef]
19. Li, B.; Gong, Y.; Gao, M.L. Failure Analysis of Hat-Stringer-Stiffened Aircraft Composite Panels under Four-Point Bending Loading. *Materials* **2022**, *15*, 2430. [CrossRef]
20. Russo, A.; Sellitto, A.; Saputo, S.; Acanfora, V.; Riccio, A. Cross-Influence between Intra-Laminar Damages and Fibre Bridging at the Skin–Stringer Interface in Stiffened Composite Panels under Compression. *Materials* **2019**, *12*, 1856. [CrossRef]
21. Zhang, J.; Zheng, J.; Zeng, J.; Yang, G.; Tong, M. Thermal and Mechanical Stress Analysis in Aircraft Hybrid Panels with Multi-Bolt Joints. *Materials* **2024**, *17*, 1872. [CrossRef]
22. Abramovich, H.; Weller, T. Repeated buckling and postbuckling behavior of laminated stringer-stiffened composite panels with and without damage. *Int. J. Struct. Stab. Dyn.* **2010**, *10*, 807–825.
23. Chang, F.K.; Chang, K.Y. A progressive damage model for laminated composites containing stress concentrations. *Compos. Mater.* **1987**, *21*, 834–855. [CrossRef]
24. Chang, K.Y.; Liu, S.; Chang, F.K. Damage tolerance of laminated composites containing an open hole and subjected to tensile loadings. *J. Compos. Mater.* **1991**, *25*, 274–301.
25. Hashin, Z. Failure criteria for unidirectional fiber composites. *J. Appl. Mech.* **1980**, *47*, 329–334.
26. Puck, A.; Schürmann, H. Failure analysis of FRP laminates by means of physically based phenomenological models. *Compos. Sci. Technol.* **1998**, *58*, 1045–1067.
27. Shokrieh, M.M.; Lessard, L.B. Progressive fatigue damage modeling of composite materials, Part I: Modeling. *J. Compos. Mater.* **2000**, *34*, 1056–1080.

28. Tserpes, K.I.; Labeas, G.; Papanikos, P.; Kermanidis, T. Strength prediction of bolted joints in graphite/epoxy composite laminates. *Compos. Part B Eng.* **2002**, *33*, 521–529.
29. Tserpes, K.I.; Papanikos, P.; Labeas, G.; Pantelakis, S. Fatigue damage accumulation and residual strength assessment of CFRP laminates. *Compos. Struct.* **2004**, *63*, 219–230. [CrossRef]

materials

MDPI

Article

Research on the Structural Design and Mechanical Properties of T800 Carbon Fiber Composite Materials in Flapping Wings

Ruojun Wang [1], Zengyan Jiang [2,*], Yuan Zhang [2], Luyao Fan [1] and Weilong Yin [1,*]

1 Center for Composite Materials and Structures, Harbin Institute of Technology, Harbin 150080, China; 23s118135@stu.hit.edu.cn (R.W.); 24s118100@stu.hit.edu.cn (L.F.)

2 Aerodynamics Research Institute, AVIC (Aviation Industry Corporation of China), Harbin 150080, China; songbinghao00001@126.com

* Correspondence: jiangzyan@avic.com (Z.J.); ywl@hit.edu.cn (W.Y.)

Abstract

Due to its superior maneuverability and concealment, the micro flapping-wing aircraft has great application prospects in both military and civilian fields. However, the development and optimization of lightweight materials have always been the key factors limiting performance enhancement. This paper designs the flapping mechanism of a single-degree-of-freedom miniature flapping wing aircraft. In this study, T800 carbon fiber composite material was used as the frame material. Three typical wing membrane materials, namely polyethylene terephthalate (PET), polyimide (PI), and non-woven kite fabric, were selected for comparative analysis. Three flapping wing configurations with different stiffness were proposed. These wings adopted carbon fiber composite material frames. The wing membrane material is bonded to the frame through a coating. Inspired by bionics, a flapping wing that mimics the membrane vein structure of insect wings is designed. By changing the type of membrane material and the distribution of carbon fiber composite materials on the wing, the stiffness of the flapping wing can be controlled, thereby affecting the mechanical properties of the flapping wing aircraft. The modal analysis of the flapping-wing structure was conducted using the finite element analysis method, and the experimental prototype was fabricated by using 3D printing technology. To evaluate the influence of different wing membrane materials on lift performance, a high-precision force measurement experimental platform was built, systematic tests were carried out, and the lift characteristics under different flapping frequencies were analyzed. Through computational modeling and experiments, it has been proven that under the same flapping wing frequency, the T800 carbon fiber composite material frame can significantly improve the stiffness and durability of the flapping wing. In addition, the selection of wing membrane materials has a significant impact on lift performance. Among the test materials, the PET wing film demonstrated excellent stability and lift performance under high-frequency conditions. This research provides crucial experimental evidence for the optimal selection of wing membrane materials for micro flapping-wing aircraft, verifies the application potential of T800 carbon fiber composite materials in micro flapping-wing aircraft, and opens up new avenues for the application of advanced composite materials in high-performance micro flapping-wing aircraft.

Keywords: micro flapping-wing aircraft; T800 carbon fiber composite material; finite element analysis; aerodynamic characteristics; 3D printing

1. Introduction

The design inspiration for flapping-wing aircraft stems from the biomimetic research on the flight mechanisms of birds and insects in nature [1]. Compared to traditional fixed-wing and rotor-wing aircraft, flapping-wing aircraft can precisely replicate the periodic flapping motion and three-dimensional twisting motion of biological wings, achieving efficient aerodynamic performance in low Reynolds number fluid environments [2]. The biomimetic structure design endows it with excellent adaptability to complex environments, enabling stable flight in confined spaces and possessing high maneuverability, low noise, and other advantages [3]. These characteristics mean that flapping-wing aircraft have significant application value in the field of micro unmanned aircraft [4,5].

Lightweight technology, as a key approach to enhancing the comprehensive performance of aircraft, optimizing energy utilization efficiency, and increasing effective carrying capacity, has always been the core of research in the aerospace field [6]. Therefore, weight is one of the important indicators for measuring the advancedness of aircraft design. Under the premise of meeting flight tasks, reducing the weight of aircraft is the eternal pursuit of designers. If lightweight design is applied to flapping-wing aircraft, it means a lighter fuselage, less fuel consumption, greater flexibility and convenience, and thus greater market prospects. Therefore, we hope to design flapping-wing aircraft with lightweight structures under the premise of meeting certain strength, stiffness, and lifespan requirements. There are mainly two ways of lightweighting: one is structural optimization [7,8], and the other is the use of stronger materials, especially advanced composite materials [9–11].

Structure is the skeleton of aerospace equipment, and for a long time, it has constituted the main object of system lightweighting. Lightweighting of material structure requires the full and reasonable application of high-performance lightweight materials (such as lightweight alloys, composite materials, foam/foam core/particle lattice materials, etc.) [12], new structural optimization design methods (topology optimization, overall optimization, etc.) [13,14] and new process technologies (additive manufacturing, composite manufacturing, etc.), achieving the optimization and lightweighting of multiple carrying performances through the rational layout and parameter optimization of materials in the structural space.

With the rapid development of biomimetic aircraft technology, flapping-wing aircraft, due to their efficient aerodynamic characteristics and biomimetic maneuverability, have become a research hotspot. Traditional metal materials are more commonly used in large aircraft, but in the field of flapping-wing aircraft, they face challenges such as heavy weight, insufficient fatigue performance, etc., in achieving high lift, low energy consumption and long lifespan. Composites, with their lightweight, high specific strength, excellent fatigue resistance and designability, provide new ideas for flapping-wing aircraft structure design and performance improvement, and have promoted significant progress in the research of micro flapping-wing aircraft. The application of composite materials in modern flapping-wing aircraft mainly lies in weight reduction, aerodynamic optimization and functional integration [15,16]. In terms of weight reduction, carbon fiber reinforced composite materials have extremely high strength and low density and are widely used [17,18]. In terms of aerodynamic optimization, resin-based composite materials, through integrated molding processes, have high precision characteristics and have broad application prospects in micro flapping-wing aircraft. International research on composite materials in flapping-wing aircraft has formed a relatively standardized integration model, mainly concentrated in research institutions in the United States, Japan and Europe. The US DARPA proposed the concept of micro aircraft in 1992, promoting the application of composite materials in micro flapping-wing aircraft [19]. The Robobees series from Harvard University employs the Smart Composite Microstructure (SCM) technology, embedding piezoelectric materials into the composite wing structure, achieving an ultra-miniature design with a 3 cm wingspan and 80 mg weight. The latest generation of Robobee X-Wing, through the optimization of the four-wing structure and lightweight design of the composite material, weighs 259 mg and has a thrust efficiency close to that of insects of the same volume [20,21]. The Defly Nimble flapping robot from Delft University of Technology uses transparent polyester film composite wing structures, achieving agile flight movements similar to those of fruit flies, including sharp turns and back flips at a speed of 25 km/h. This design fully demonstrates the advantages of composite materials in achieving complex kinematics [22].

In the field of micro-bionic flapping-wing aircraft, polymer materials, due to their low modulus characteristics, can meet the requirements for the high-frequency reciprocating motion of the flapping wings. Polymer materials also have excellent specific strength and fatigue resistance, which can ensure the structural stability of the mechanism under long-term cyclic loading. Therefore, polymer materials can be used as the wing surface framework to achieve lightweight configuration design, significantly reducing energy loss during the flight process. Their excellent molding processability is applied in 3D printing, injection molding and other processes, enabling the high-precision fabrication of complex bionic structures [23,24]. Advanced composite materials formed by combining high-performance reinforcing elements such as carbon fibers and glass fibers with polymer matrices can effectively meet the dual performance requirements of flapping-wing structures for flexible deformation ability and structural stiffness. The stiffness of the wing shape (including rigid wings and flexible wings) of flapping-wing aircraft has a decisive impact on its aerodynamic performance. In recent years, scholars at home and abroad have conducted in-depth research on the energy efficiency and biomimetic optimization of these two types of wing shapes, with the aim of breaking through the performance limitations of traditional aircraft [25–27]. Through structural optimization design guided by bionics principles, such composite materials can further improve the aerodynamic efficiency of the flapping system and significantly enhance key performance parameters such as flapping frequency. Studies have shown that this material structure integrated design method not only solves the aerodynamic efficiency bottleneck problem of traditional aircraft in low Reynolds number conditions but also provides a new technical path for the performance improvement of micro aircraft. Especially in scenarios that require both lightweighting and structural strength, composite flapping-wing aircraft demonstrate irreplaceable technical advantages [28–31].

2. Materials and Methods

Single-Degree-of-Freedom Fluttering Mechanism

To investigate the influence of different wing membrane materials on the aerodynamic characteristics of bionic aircraft, this study designed a single-degree-of-freedom flapping mechanism, as illustrated in Figure 1. Based on this mechanism, experimental research was conducted. This mechanism achieves flapping motion through a single driving source, which significantly reduces mechanical complexity and enhances the system's power-to-weight ratio, thereby fulfilling the lightweight design requirements for bionic aircraft. The simplified structure effectively avoids coordination control and motion coupling issues inherent in multi-degree-of-freedom systems, enabling rapid construction and debugging while minimizing processing costs, assembly errors, and potential failure points. Consequently, the stability and repeatability of the experimental system are improved. Moreover, the single-degree-of-freedom design facilitates precise control over key motion parameters such as flapping frequency and amplitude.

In this study, the simplified modeling method was adopted to conduct static simulation analysis on the gear transmission mechanism. Based on the local effect assumption of saint venant principle [32], the influence of secondary structural components far from the load action area on the gear contact stress distribution can be ignored. Therefore, when conducting strength checks on a single-degree-of-freedom fluttering mechanism, the connecting rod structure can be omitted, and the analysis object can be simplified to a transmission system composed of three meshing gears. This modeling simplification strategy can effectively reduce the model complexity. On the premise of ensuring the calculation accuracy of key mechanical parameters such as tooth surface contact stress and tooth root bending stress [33], it realizes the reasonable allocation of computing resources to the mesh refinement of the gear meshing area.

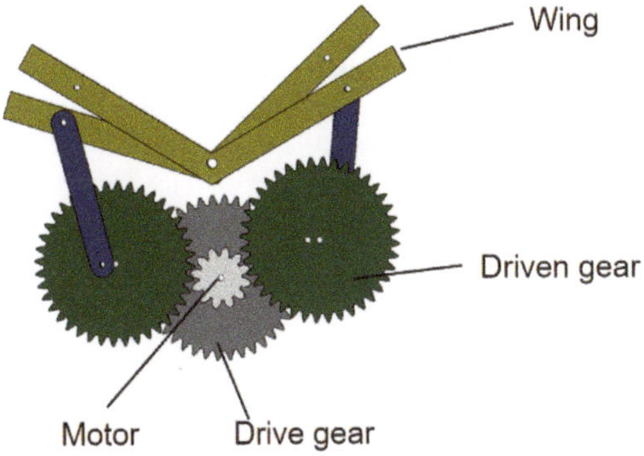

Figure 1. Schematic diagram of single-degree-of-freedom flapping mechanism.

In order to evaluate whether the use of nylon material meets the requirements of gear strength, a finite element analysis model was established based on the designed single-degree-of-freedom flapping mechanism. As shown in the figure, in this study, the numerical model of the gear meshing part of the flapping mechanism was constructed by using the Abaqus/Standard (2020) commercial finite element analysis software. The geometric nonlinear effect (based on the finite deformation theory) was considered in the analysis process. The model parameters are set as follows: The module of the driving gear connected to the drive motor is 0.3 and the number of teeth is 9. The modules of the two driven gears are both 0.3 and the number of teeth is both 40. The model was meshed by using the C3D8R element type, and the strength of the mechanism was checked through stress analysis. The simulation analysis results show that, as shown in Figure 2, the maximum stress value of the flinging mechanism is 166.1 Pa, and its position is located in the connection area between the driving gear and the motor. This stress value is significantly lower than the yield strength limit of the nylon material. From this, it can be determined that the selected nylon material meets the design requirements in terms of strength performance and has reliable mechanical load-bearing capacity.

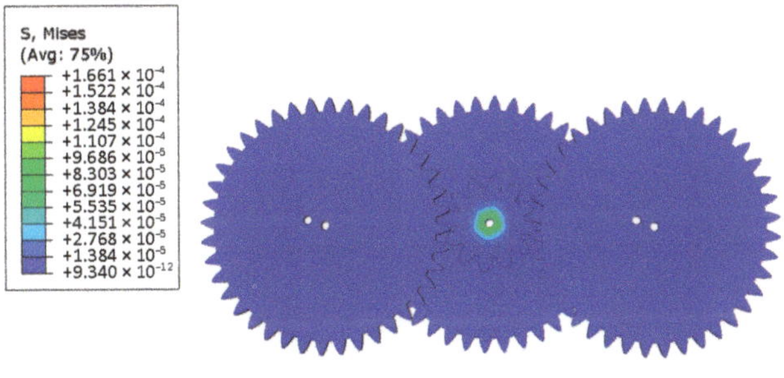

Figure 2. Cloud diagram of stress verification for gear mechanism.

3. Preparation of Flapping Wing Prototype

3.1. Flapping Wing Shape Design

Based on the above finite element analysis results, an improved design was carried out based on the wing membrane and wing vein. For the lightweight design of the wing, the wing shape was first designed according to the aerodynamic performance. The rigid front is composed of carbon fiber rods, and the flexible wing is composed of polyester film, as shown in Figure 3. The trailing edge of the wing is fixed on the fuselage to ensure that the flexible material of the wing surface can undergo significant deformation during the flapping process of the wing. To a certain extent, the torsional movement of the wings is achieved, enabling the prototype to obtain better aerodynamic performance. Based on the wing design of "wing membrane plus wing vein", carbon fiber rods are used to wrap the entire wing membrane around it to enhance rigidity. The designed mechanism is connected to the flapping mechanism. With this scheme, the wing rigidity increases.

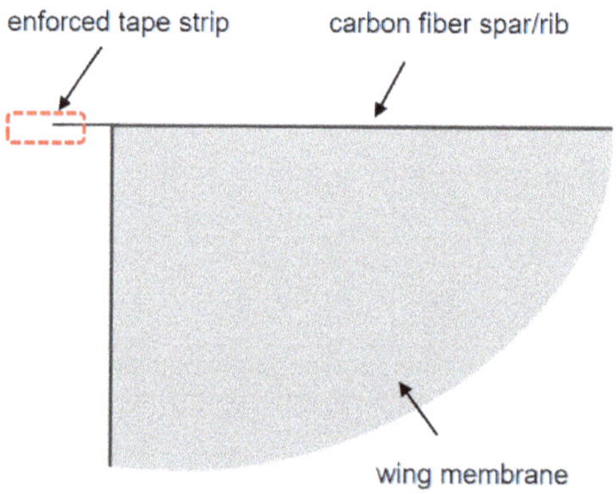

Figure 3. Semi-enveloped wing configuration.

In order to imitate the morphological and functional characteristics of natural biological wings more accurately, this paper proposes a bionic airfoil design scheme based on the composite structure of "wing membrane-wing vein", as shown in Figure 4. In terms of the selection of wing surface materials, polyester film with a thickness of 0.0125 mm was used as the main load-bearing surface. The ultra-thin polyester film not only provides a certain strength but also minimizes the weight of the wing itself to the greatest extent. To enhance the aerodynamic performance of the airfoil mechanism, carbon fiber reinforced composite materials with diameters of 1 mm are used to support the wing surfaces at the leading edge and the wing root of the wing, as shown in Figure 5. The passive pitch motion during the flapping process is guaranteed through the sleeve type non-fixed connection method. This design effectively simulates the adaptive deformation characteristics of natural wings during the flapping process.

This study conducted a modal analysis of the wing structure. During the modeling process, the geometric shape of the wing was simplified to an elliptical wing surface structure. Three different wing structures were designed, as shown in Figure 6. The wing vein structure was simulated using T800 carbon fiber rods, and its material parameters were as follows: the density was 1.8 g/cm^3, the elastic modulus was 300 GPa, and the Poisson's ratio was 0.3. The finned membrane structure was simulated using PET film material, and its material parameters were density 1.323 g/cm^3, elastic modulus 28 GPa, and Poisson's ratio 0.3. In the process of finite Element modeling, Shell elements are selected for discretization processing, and no additional constraint conditions need to be applied.

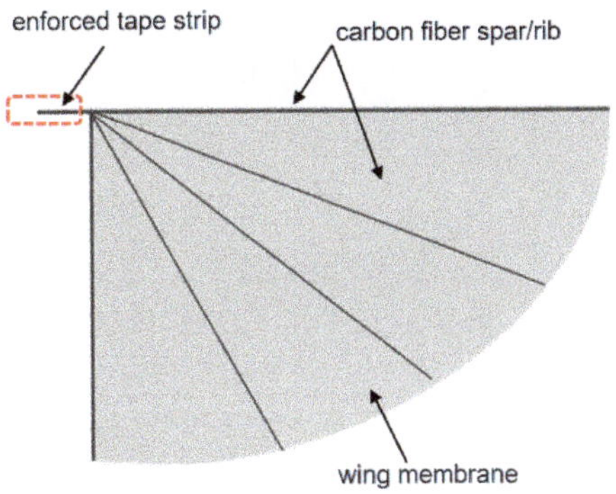

Figure 4. Schematic diagram of the wing membrane-wing vein structure.

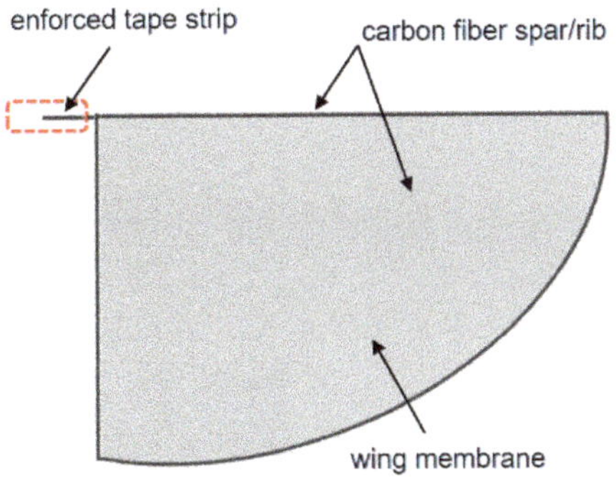

Figure 5. Fully enclosed flapping-wing structure.

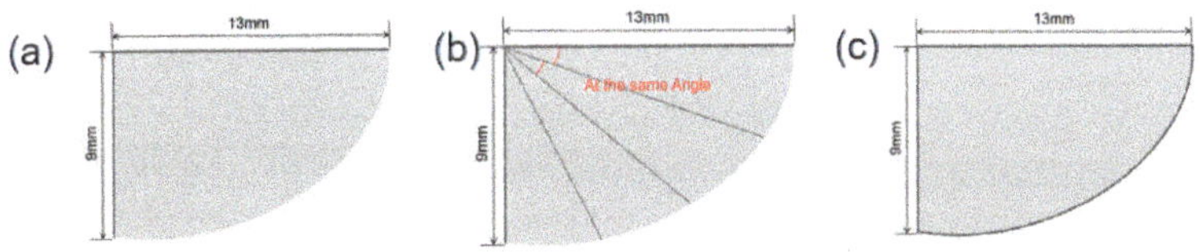

Figure 6. (**a**) Wing configuration 1 (**b**) wing configuration 2 (**c**) wing configuration 3.

As can be seen from Figure 7, the first six natural frequencies of the fixed two constrained elliptical wings on both sides are 0.11386 Hz, 0.26436 Hz, 0.40122 Hz, 0.48529 Hz, 0.69106 Hz, and 0.77702 Hz, respectively. The simulation results of the first six vibration modes of the elliptical wing are shown in the following figure. It can be seen that the vibrations all occur at the posterior edge of the elliptical wings. Since the wing veins at the posterior edge have no wing veins and the bending stiffness is very low, this area is prone to inducing vibrations.

Imitating the wing membrane-wing vein structure of insect wings, T800 carbon fiber rods are introduced into the model as wing vein materials to enhance rigidity. As can be seen from Figure 8, the natural frequencies of the first six orders of the optimized structure have significantly increased, which are 1.1551 Hz, 1.3692 Hz, 1.5081 Hz, 1.7587 Hz, 2.1190 Hz and 2.2870 Hz, respectively. The mode analysis shows that, although the vibration still mainly occurs at the trailing edge of the wing, the vibration area is significantly reduced, indicating that the introduction of the wing veins effectively enhances the local stiffness and suppresses the overall vibration response of the structure.

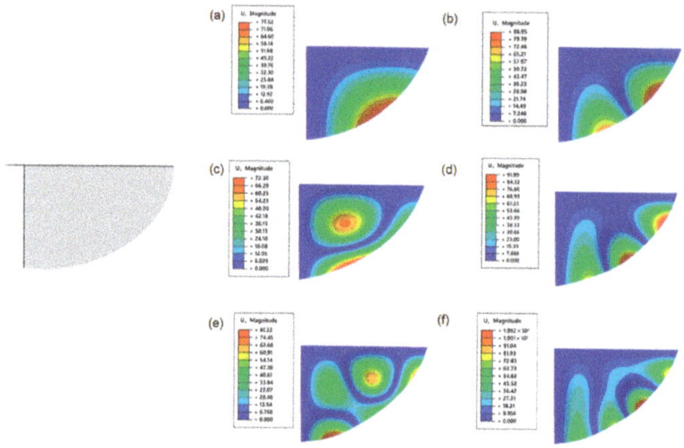

Figure 7. Semi-enveloped wing configuration—the first six vibration modes of the elliptical wings (**a**) C1-first-order mode (vibration in the middle section of the wing). (**b**) C1-The second-order mode (the vibration intensity in the wings has increased). (**c**) C1-The third-order mode (the vibration area spreads towards the middle of the wing). (**d**) C1-The fourth-order mode (coupling vibration occurs). (**e**) C1-The fifth-order mode (the coupling vibration intensity has increased). (**f**) C1-The sixth-order mode (coupled vibrations are obvious).

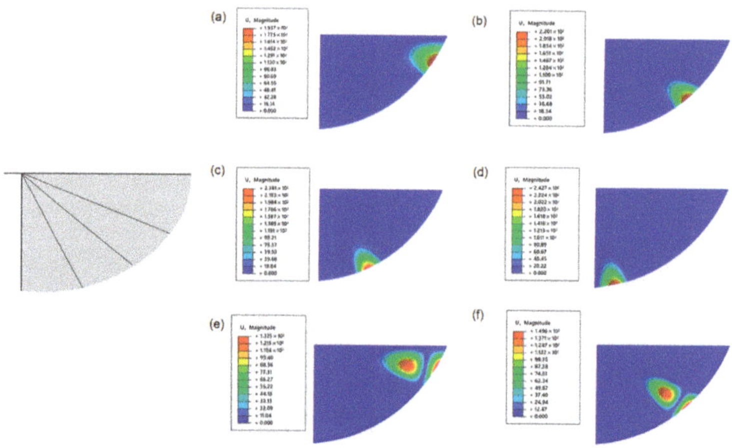

Figure 8. Schematic diagram of the wing membrane-wing vein structure—the first six vibration modes of the elliptical wings. (**a**) C2-first-order mode (low-order vibration in the aft edge area of the wing tip). (**b**) C2-The second-order mode (vibration at the middle part of the wing's trailing edge). (**c**) C2-The third-order mode (the vibration spreads from the rear edge root of the wing). (**d**) C2-The fourth-order mode (vibration in the wing root area). (**e**) C2-The fifth-order mode (Wingtip coupling vibration). (**f**) C2-The sixth-order mode (coupled vibration of the wing trailing edge).

As can be seen from Figure 9, the entire area around the elliptical wing was fixed, and the first six natural frequencies were increased to 0.40086 Hz, 0.67604 Hz, 0.89124 Hz, 1.0339 Hz, 1.2965 Hz, and 1.4711 Hz. It can be seen that the vibration concentration area shifted to the middle of the wing. The reason is that no wing vein support was set in this area. The relatively low structural stiffness makes it a sensitive area of the vibration-dominated mode.

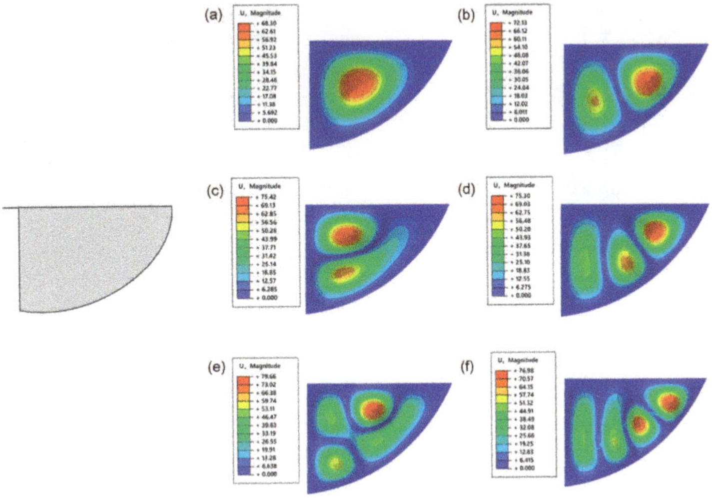

Figure 9. Fully enclosed flapping-wing structure—the first six vibration modes of the elliptical wings. (**a**) C3-first-order mode (vibration in the middle part of the wing). (**b**) C3-The second-order mode (coupled vibration occurs in the middle part of the wing). (**c**) C3-The third-order mode (the vibration intensity in the middle part of the wing has increased). (**d**) C3-The fourth-order mode (the vibration in the middle part of the wing is obvious). (**e**) C3-The fifth-order mode (coupled vibration in the middle section of the wing) (**f**) C3-The sixth-order mode (high-order coupled vibration in the middle section of the wing).

This study shows that the introduction of wing vein materials can significantly enhance the overall stiffness of the structure, thereby significantly increasing the natural frequency. In flapping-wing structures, the low-stiffness areas (such as the trailing edge or middle part without wing vein support) are more prone to form vibration concentration areas due to their weaker bending resistance. By optimizing the spatial arrangement of high-stiffness materials (such as wing veins), effective regulation of vibration modes can be achieved, thereby reducing the vibration amplitude and its influence range.

3.2. Selection of Flapping Wing Materials

The wing materials of flexible flapping-wing aircraft typically include two types: wing ribs for enhancing stiffness and lightweight wing membranes. The selection of these materials should aim to achieve lightweight structure and functional synergy while meeting mechanical and aerodynamic performance requirements. The wing is a movable part. The quality of it will directly affect the magnitude of its moment of inertia. For moving parts, we should reduce their weight to decrease the inertia of movement. Density is the first factor we need to consider. Apart from the drive power supply, circuit board, coil and other components, the fuselage frame is the component with the largest number of parts and also the one that accounts for the largest proportion of weight. Wing ribs mainly serve as structural supports, and must have high specific strength and specific stiffness to effectively reduce weight while maintaining rigidity, thereby reducing the rotational inertia of the wings during flight and improving maneuverability. Therefore, in order to ensure that the mass of the entire flapping wing device is as small as possible, the density of the selected frame material should also be as low as possible. In order to increase the rigidity of the wing, the wing vein structure needs to be introduced, which should have a certain strength and be able to withstand deformation. Selecting materials with a lower density to manufacture the wing vein components can achieve a smaller wing mass. T800 carbon fiber composite material features low density (1.6 g/cm^3), extremely high tensile strength and rigidity, which can significantly reduce the weight of the structure, while enhancing the anti-deformation capacity and durability of the wings and structural frames, thereby improving the structural stability of the aircraft [34–36]. In addition, T800 material has good designability and its performance can be further optimized by adjusting the arrangement of fibers. With its outstanding mechanical properties and lightweight characteristics, T800 carbon fiber composite material is highly suitable for application in the design of micro aircraft (MAVs) that have high requirements for performance and weight. Therefore, in this paper, T800 carbon fiber composite material is selected as the wing frame material to meet the requirements of structural strength and lightweight. Considering the above points comprehensively, T800 carbon fiber with a diameter of 1mm is selected as the wing frame structure.

In the structural design of this project, the wing membrane is adhered to the wing ribs and directly participates in the transmission of aerodynamic loads. The material of the wing membrane must have good flexibility and airtightness to withstand periodic large deformations of flapping motion and maintain the stability of lift output. To ensure the long-term reliable operation of the overall structure under complex conditions, the selection of the wing membrane material should meet the following requirements: first, low density to reduce wing surface mass and optimize dynamic response; second, excellent adhesion to enhance the connection stability between the membrane material and the framework; third, good flexibility to adapt to elastic deformation during flight; fourth, shape retention ability. To more closely resemble the shape of an insect's wing, the wing surface should be as close as possible to the biomimetic shape. Therefore, after production, the wing membrane material should not be prone to deformation to maintain its expected aerodynamic shape. These performance requirements collectively form the design basis for the selection of flexible wing materials.

When designing flap-wing aircraft, PET film materials, polyimide film materials and non-woven kite cloth are mainly selected for experimental comparison, with the aim of choosing wing film materials with good performance and light weight. The surface energy of PET film is relatively low, and it is hydrophobic and chemically inert [37]. The weak Lewis acid-base properties of the PET film surface imply that the interfacial interaction force between it and other materials (such as adhesives, composite layers) is relatively small. Polyimide films have a relatively high surface energy, which is conducive to the adhesion between materials and the bonding of interfaces [38], when micro flapping wings are combined with adhesives or composite materials, the Lewis acid-base properties of the polyimide film surface are conducive to the formation of a stable interface layer. Kite cloth film, as a coating or composite material, has a relatively rough surface [39]. These differences have a significant impact on the mechanical properties and interface behavior of the wing membrane. Therefore, these three materials were selected for the design of the wing membrane to explore the influence of materials with different properties on the mechanical properties of the flapping mechanism. The three types of wing membrane materials are shown in Figure 10, and their material parameters are provided in Table 1.

Figure 10. Three types of wing membrane materials.

Table 1. Properties of three types of wing membrane materials.

Material Type	Density (g/cm^3)	Tensile Strength (MPa)	Elastic Modulus (GPa)	Elongation at Break (%)	Surface Energy (mJ/m^3)
PET	1.38	50–80	2–4	50–150	40–45
polyimide	1.42	150–300	2.5–4	20–50	40–50
kite fabric	0.9–1.2	100–300	–	10–30	–

This study adopted elliptical wings as the experimental objects. Although automated processing technologies such as laser cutting can achieve high-precision and high-efficiency cutting of wing films, due to the high equipment cost, and the relatively simple processing technology and low manufacturing cost of the flapping wing structure itself, automated processing methods are usually not adopted to handle wing film materials in actual production. Therefore, in this study, the manual cutting method was chosen and a unified template was used to prepare elliptical wing membrane samples. Three kinds of wing membrane materials were selected for the experiment: 0.0125 mm thick PET film, 0.05 mm thick polyimide film and non-woven kite fabric. The samples of three bionic wing membrane materials after cutting and processing are shown in the Figure 11.

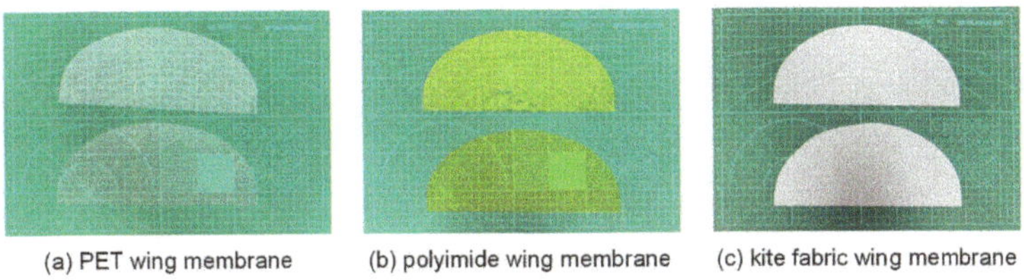

Figure 11. Three kinds of wing membrane cutting and preparation.

Three types of wing membrane materials were cut and assembled, respectively, according to the designed configuration. The assembled wing membrane configuration is shown in the Figure 12.

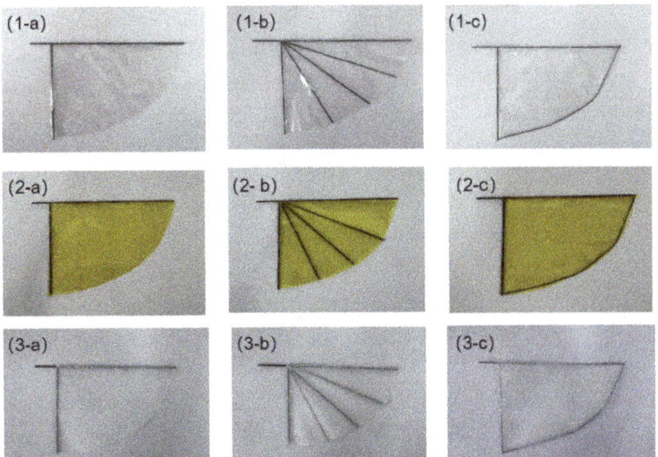

Figure 12. Preparation of three different configurations of wing membranes. (**1-a**) the assembled wing membrane-c1-PET (**1-b**) the assembled wing membrane -c2-PET (**1-c**) the assembled wing membrane-c3-PET (**2-a**) the assembled wing membrane- c1-PI (**2-b**) the assembled wing membrane- c2-PI (**2-c**) the assembled wing membrane -c3-PI (**3-a**) the assembled wing membrane -c1-Kite (**3-b**) the assembled wing membrane- c2-Kite (**3-c**) the assembled wing membrane- c3-Kite.

3.3. Pulsating Mechanism Printing

The single-degree-of-freedom flutter mechanism designed in this paper is printed using DLP light processing technology. The 3D printer used is the Rayshape P400 (Raise3D, Shanghai, China), which has a high projection resolution of 3840 × 2160 pixels and a printing speed of 15 mm per hour. Resin material is selected for printing. The printed single-degree-of-freedom components are assembled. After assembly, the pre-cut wing membrane material is adhered to the surface of the connection rod of the flutter mechanism with glue. The following Figure 13 shows the prototype of the flutter mechanism used for experimental testing.

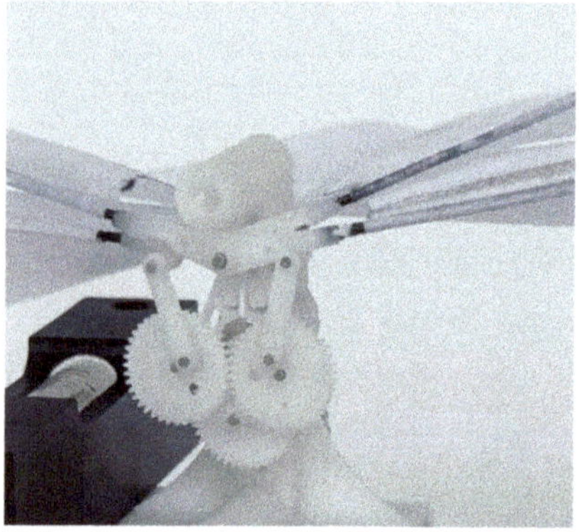

Figure 13. Diagram of the experimental prototype.

4. Experiment

4.1. Construction of the Force Measurement Experiment Platform

After setting up the experimental platform, data collection was carried out on the force measurement platform of the single-degree-of-freedom flinging mechanism that had already been established. The experimental situation is as shown in the Figure 14. The voltage value range is 0–4 V. According to the control throttle, it is divided into four gears to control the voltage change.

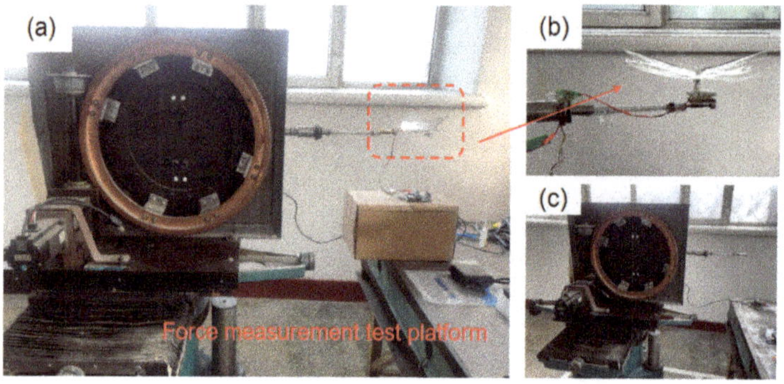

Figure 14. Single-degree-of-freedom force measurement experimental platform (**a**) Schematic diagram of the force measurement experiment process (**b**) Installation diagram of the oscillation mechanism (**c**) Force measurement experimental platform v.

To achieve precise measurement of the flapping frequency of the single-degree-of-freedom flapping mechanism, a Hall sensor assembly is installed on the mechanism. This sensor is capable of sensing the periodic displacement changes during the movement of the flapping wing, thereby achieving real-time monitoring of the flapping period and calculating the corresponding flapping frequency accordingly. This non-contact measurement method can effectively enhance the stability and reliability of experimental data. The fully built experimental test platform is shown in the Figure 15.

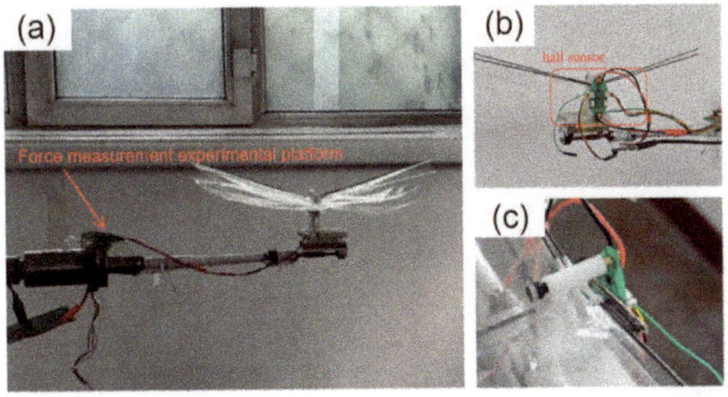

Figure 15. Single-degree-of-freedom Hall sensor experiment. (**a**) The fluttering mechanism installed on the force measurement experimental platform. (**b**) Installation diagram of Hall sensor. (**c**) Local schematic diagram of Hall sensor position.

In order to study the influence of different wing membrane materials and different stiffness configurations on the mechanical properties of the single-degree-of-freedom flinging mechanism, under the premise of ensuring the mechanism structure and driving power supply remain unchanged, three different types of wing membrane materials were selected for replacement, and comparative experiments were carried out, respectively. Through comparative experiments, the influence of wing membrane materials on performance indicators such as lift and frequency was evaluated. The process of changing the wing membrane material and the experiment under different configurations is shown in the Figure 16.

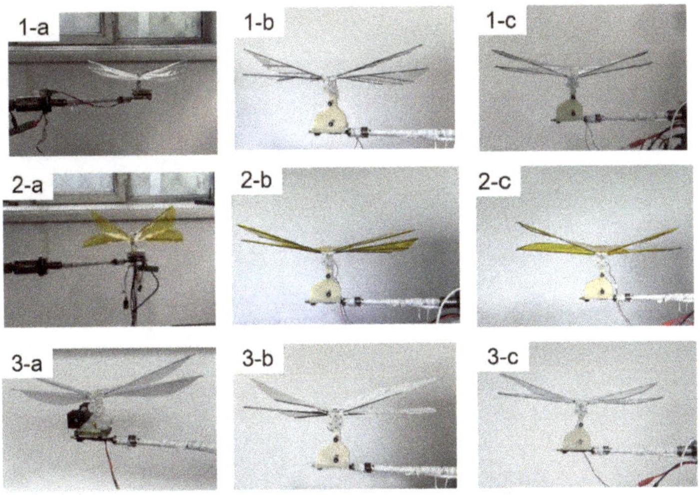

Figure 16. Comparative experiments of three types of wing membrane materials.(**1-a**) c1-PET experimental schematic diagram (**1-b**) c2-PET experimental schematic diagram (**1-c**) c3-PET experimental schematic diagram (**2-a**) c1-PI experimental schematic diagram (**2-b**) c2-PI experimental schematic diagram (**2-c**) c3-PI experimental schematic diagram (**3-a**) c1-Kite experimental schematic diagram (**3-b**) c2-Kite experimental schematic diagram (**3-c**) c3-Kite experimental schematic diagram.

4.2. Experimental Data Analysis

To explore the influence of wing membrane materials and airfoil configurations on the mechanical properties of the flapping mechanism, force measurement and frequency measurement experiments were conducted on three combinations of wing membrane materials and three configurations, respectively. To enhance the reliability and representativeness of the experimental results, each group of experiments was independently measured three times, and the average value of the obtained data was calculated to reduce the influence of accidental errors. The experimental data are shown in the following table. By conducting post-processing on the experimental data and drawing bar graphs, it is convenient to compare the influence of different factors on mechanical properties more intuitively. As shown in Figure 17, it presents the variation laws of lift and flutter frequencies corresponding to different wing membrane materials under the same configuration conditions; Figure 18 shows the variation laws of lift and frequency corresponding to different flapping wing configurations under the same wing membrane material conditions.

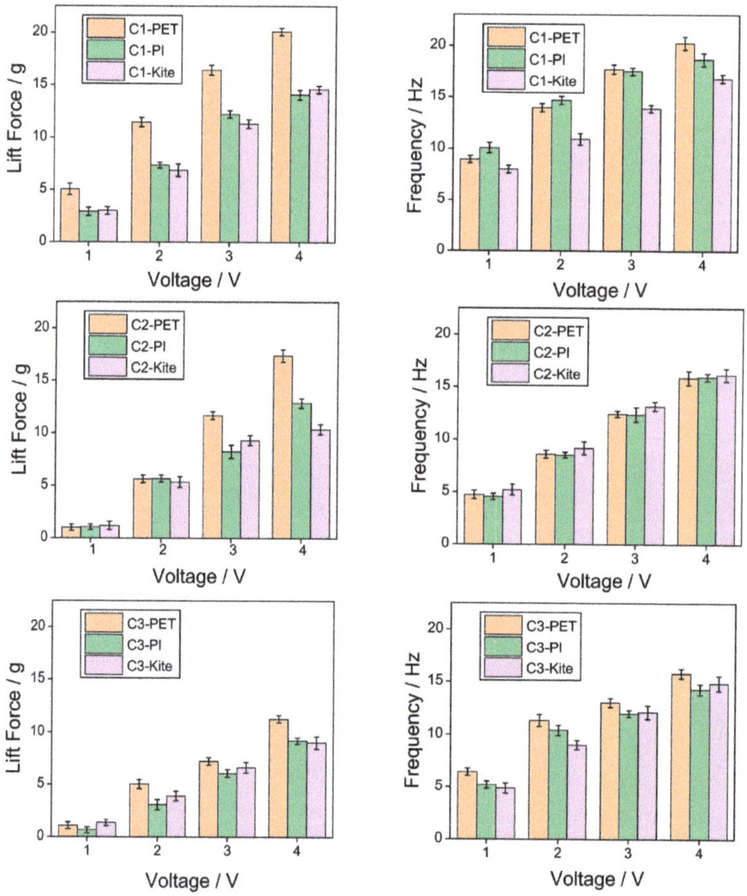

Figure 17. The effects of three different materials on lift and frequency.

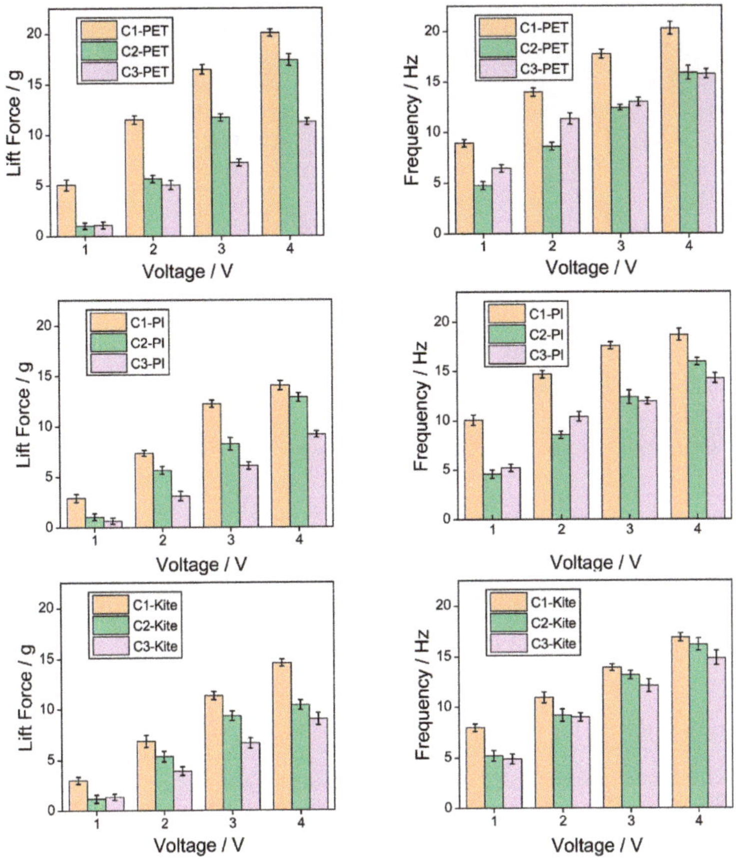

Figure 18. The effects of three different configurations on lift and frequency.

From the data in the above Tables 2 and 3, it can be seen that the highest flapping frequency of 20.27 Hz and the flapping lift of 20.07 g both originated from configuration 1 of the PET film material. It can be seen from Figure 17 that under the same configuration, the lift of the pet film is significantly higher than that of the polyimide film and the kite cloth film. From Figure 18, it can be seen that under low-voltage drive, the lighter structure can generate a relatively large lift force with a smaller voltage input. Although the designs of configuration 2 and configuration 3 offer strong structural stiffness, the additional weight offsets part of the lift effect, resulting in their lift being much lower than that of configuration 1.

Table 2. Lift data from three-wing membrane three-configuration force measurement experiments.

Voltage	C1-PET	C2-PET	C3-PET	C1-PI	C2-PI	C3-PI	C1-Kite	C2-Kite	C3-Kite
1 V	5.07 g	1.01 g	1.09 g	2.91 g	1.05 g	0.66 g	3.02 g	1.19 g	1.38 g
2 V	11.47 g	5.66 g	5.05 g	7.36 g	5.69 g	3.11 g	6.87 g	5.36 g	3.92 g
3 V	16.48 g	11.69 g	7.24 g	12.25 g	8.25 g	6.09 g	11.32 g	9.32 g	6.66 g
4 V	20.07 g	17.40 g	11.25 g	14.04 g	12.87 g	9.18 g	14.77 g	10.39 g	9.01 g

Table 3. Frequency data from three-wing membrane three-configuration force measurement experiments.

Voltage	C1-PET	C2-PET	C3-PET	C1-PI	C2-PI	C3-PI	C1-Kite	C2-Kite	C3-Kite
1 V	8.94 Hz	4.75 Hz	6.44 Hz	10.07 Hz	4.56 Hz	5.21 Hz	7.97 Hz	5.21 Hz	4.88 Hz
2 V	13.98 Hz	8.59 Hz	11.32 Hz	14.69 Hz	8.53 Hz	10.41 Hz	10.93 Hz	9.17 Hz	8.98 Hz
3 V	17.76 Hz	12.43 Hz	13.01 Hz	17.56 Hz	12.37 Hz	11.95 Hz	13.90 Hz	13.15 Hz	12.11 Hz
4 V	20.27 Hz	15.88 Hz	15.77 Hz	18.66 Hz	15.95 Hz	14.28 Hz	16.87 Hz	16.17 Hz	14.84 Hz

In this study, three distinct wing configurations, each paired with three different membrane materials, were systematically designed and evaluated. Force measurement experiments were conducted on the established force measurement platform to evaluate the performance indicators of lift and frequency. Among these designs, the C1-PET configuration, which exhibited the best lift performance, was selected for a comparative analysis with two well-established bio-inspired micro air vehicles: Harvard's RoboBee and DelFly Nimble from Delft University of Technology. The results are summarized in Table 4. As shown in the table, the proposed C1-PET design achieves a maximum lift of 20.07 g, with a total system weight of 5.3 g, resulting in a lift-to-weight ratio of approximately 3.79, the performance is between that of RoboBee [20,21] and DelFly Nimble [22] products. Furthermore, the proposed flapping-wing mechanism operates at a moderate flapping frequency of 20.27 Hz, eliminating the need for high-frequency actuation. This not only helps to reduce mechanical wear but also simplifies control complexity.

Table 4. The performance comparison of the flapping mechanism designed in this paper with Robobee and Defly.

Voltage	Voltage (V)	Lift (g)	Weight (g)	Lift to Weight	Flapping Frequency (Hz)
C1-PET	4.3	20.07	5.3	3.79	20.27
Robobee	100	0.7	0.175	4	120
Defly Nimble	8.4	45	29	1.55	17

Overall, the C1-PET configuration is competitive in terms of aerodynamic efficiency, structural simplicity and practicality, demonstrating the application potential of the flapping mechanism designed in this paper in the micro aircraft market.

5. Conclusions

In order to investigate the influence of T800 composite materials and different wing membrane materials on the mechanical properties of the flapping mechanism, this paper designs a single-degree-of-freedom flapping mechanism. Based on the different distribution patterns of T800 carbon fiber rods on the wing membrane surface, three wing structures are developed. Mechanical property comparison experiments are conducted by combining three different wing membrane materials, and the following conclusions can be drawn.

1. By introducing the T800 carbon fiber composite material-designed biomimetic "wing membrane-wing vein" structure in specific areas of the wing membrane, the local structural stiffness was effectively enhanced, and the structural vibration was suppressed, thereby improving the stability of flapping-wing flight.
2. Under low-voltage driving conditions, although the semi-covered structure has lower stiffness, its lift performance is superior to the other two structures, demonstrating good lightweighting and lift balance, and having certain potential for engineering applications.
3. Under the same configuration, the lift performance of the wing surface using PET membrane material is significantly better than that of polyimide membrane and kite fabric materials.

4. The mechanical property data of the three types of wing membrane materials (C2 and C3) indicate that in engineering practice, a solution can be adopted that combines the artificial wing membrane-wing vein structure with lightweight and high-performance wing membrane materials to enhance the aerodynamic performance and structural reliability of micro flapping-wing aircraft.

Author Contributions: Conceptualization, R.W. and L.F.; methodology, Y.Z.; software, R.W.; validation, Y.Z. and Z.J.; formal analysis, Z.J.; investigation, R.W.; resources, L.F.; data curation, W.Y.; writing—original draft preparation, R.W.; writing—review and editing, Y.Z.; visualization, Z.J.; supervision, W.Y.; project administration, L.F.; funding acquisition, W.Y. All authors have read and agreed to the published version of the manuscript.

Funding: This research received no external funding.

Institutional Review Board Statement: Not applicable.

Informed Consent Statement: Not applicable.

Data Availability Statement: The datasets presented in this article are not readily available because these data are part of the ongoing research. Requests to access the datasets should be directed to corresponding authors.

Conflicts of Interest: Authors Zengyan Jiang and Yuan Zhang were employed by the company Aviation Industry Corporation of China. The remaining authors declare that the research was conducted in the absence of any commercial or financial relationships that could be construed as a potential conflict of interest.

References

1. Lentink, D. Biomimetics: Flying like a fly. *Nature* **2013**, *498*, 306–307. [CrossRef] [PubMed]
2. Ward, T.A.; Fearday, C.J.; Salami, E.; Binti Soin, N. A bibliometric review of progress in micro air vehicle research. *Int. J. Micro Air Veh.* **2017**, *9*, 146–165. [CrossRef]
3. Chen, C.; Zhang, T. A Review of Design and Fabrication of the Bionic Flapping Wing Micro Air Vehicles. *Micromachines* **2019**, *10*, 144. [CrossRef] [PubMed]
4. Mohsan, S.A.H.; Othman, N.Q.H.; Khan, M.A.; Amjad, H.; Zywiolek, J. A Comprehensive Review of Micro UAV Charging Techniques. *Micromachines* **2022**, *13*, 977. [CrossRef] [PubMed]
5. Dai, M.; Wu, R.; Ye, M.; Gao, K.; Chen, B.; Tao, X.; Fan, Z. The Structural Design and Optimization of a Novel Independently Driven Bionic Ornithopter. *Biomimetics* **2025**, *10*, 401. [CrossRef] [PubMed]
6. Zhu, L.; Li, N.; Childs, P.R.N. Light-weighting in aerospace component and system design. *Propuls. Power Res.* **2018**, *7*, 103–119. [CrossRef]
7. Al-Haddad, L.A.; Jaber, A.A.; Giernacki, W.; Khan, Z.H.; Ali, K.M.; Tawafik, M.A.; Humaidi, A.J. Quadcopter Unmanned Aerial Vehicle Structural Design Using an Integrated Approach of Topology Optimization and Additive Manufacturing. *Designs* **2024**, *8*, 58. [CrossRef]
8. Shen, Z.; Yin, Y.; Zhang, C.; Li, T.; Fu, H. Optimization of Composite Wheel Fender for Manned Lunar Vehicle. *J. Phys. Conf. Ser.* **2024**, *2882*, 012013. [CrossRef]
9. Kumar, D.; Mohite, P.M.; Kamle, S. Dragonfly Inspired Nanocomposite Flapping Wing for Micro Air Vehicles. *J. Bionic Eng.* **2019**, *16*, 894–903. [CrossRef]
10. Yuan, S.; Li, S.; Zhu, J.; Tang, Y. Additive manufacturing of polymeric composites from material processing to structural design. *Compos. Part B Eng.* **2021**, *219*, 108903. [CrossRef]
11. Shema, A.; Yu, C.; Wang, F. Fibrous Orientation Effects on the Deformation and Stress of an Insect-like Flapping-wing Micro Aerial Vehicle Based on ACP. *IOP Conf. Ser. Mater. Sci. Eng.* **2019**, *585*, 012060. [CrossRef]
12. Potes, F.C.; Silva, J.M.; Gamboa, P.V. Development and characterization of a natural lightweight composite solution for aircraft structural applications. *Compos. Struct.* **2016**, *136*, 430–440. [CrossRef]
13. Chen, L.; Zhang, Y.; Chen, Z.; Xu, J.; Wu, J. Topology optimization in lightweight design of a 3D-printed flapping-wing micro aerial vehicle. *Chin. J. Aeronaut.* **2020**, *33*, 3206–3219. [CrossRef]
14. Van Truong, T.; Nguyen, Q.-V.; Luong, H.T.; Nguyen, Q.S.; Au, T.K.L. Experimental and numerical analysis of an optimized flapping wing mechanism for flapping wing robots. *Eng. Res. Express* **2025**, *7*, 035506. [CrossRef]

15. Galiński, C.; Żbikowski, R. Materials challenges in the design of an insect-like flapping wing mechanism based on a four-bar linkage. *Mater. Des.* **2007**, *28*, 783–796. [CrossRef]
16. Govindarajan, G.; Sharma, R. Experimental investigation on a flapping beam with smart material actuation for underwater application. *Mech. Adv. Mater. Struct.* **2019**, *28*, 1020–1034. [CrossRef]
17. LaCroix, B.W.; Ifju, P.G. Aeroelastic Model for Macrofiber Composite Actuators on Micro Air Vehicles. *J. Aircr.* **2017**, *54*, 199–208. [CrossRef]
18. Singh, S.; Ali, T.; Zuber, M.; Basri, A.A.; Mazlan, N.; Hamidon, M.N.; Basri, E.I.; Ahmad, K.A.; Hegde, N.T.; Vaz, A.C. Micro aerial vehicle flapping actuation: An experimental analysis of crank and sliding lever mechanisms. *Eng. Res. Express* **2025**, *7*, 015511. [CrossRef]
19. Han, J.; Hui, Z.; Tian, F.; Chen, G. Review on bio-inspired flight systems and bionic aerodynamics. *Chin. J. Aeronaut.* **2021**, *34*, 170–186. [CrossRef]
20. Wood, R.; Nagpal, R.; Wei, G.Y. Flight of the robobees. *Sci. Am.* **2013**, *308*, 60–65. [CrossRef] [PubMed]
21. Vaughan, O. RoboBee breaks free. *Nat. Electron.* **2019**, *2*, 265. [CrossRef]
22. Karasek, M.; Muijres, F.T.; De Wagter, C.; Remes, B.D.W.; de Croon, G. A tailless aerial robotic flapper reveals that flies use torque coupling in rapid banked turns. *Science* **2018**, *361*, 1089–1094. [CrossRef] [PubMed]
23. Rashmikant; Ishihara, D.; Suetsugu, R.; Ramegowda, P.C. One-wing polymer micromachined transmission for insect-inspired flapping wing nano air vehicles. *Eng. Res. Express* **2021**, *3*, 045006. [CrossRef]
24. Quadrini, F.; Bellisario, D.; Iorio, L.; Santo, L. Shape memory polymer composites by molding aeronautical prepregs with shape memory polymer interlayers. *Mater. Res. Express* **2019**, *6*, 115711. [CrossRef]
25. Li, Z.; Cui, L.; Wang, H.; Zhang, F.; Liu, Z.; Wang, G. Insect-inspired passive wing collision recovery in flapping wing microrobots. *Bioinspir. Biomim.* **2025**, *20*, 026004. [CrossRef] [PubMed]
26. Zhang, L.; Zhu, Y.; Yu, S.; Duan, S.; Zhang, P. Design and flight verification of passive wing folding for bionic flapping-wing aerial vehicle based on flexible joints. *Eng. Res. Express* **2025**, *7*, 035508. [CrossRef]
27. Yeh, S.I.; Chiang, C.H. The Influence of Wing Membrane Elasticity on Aerodynamics in a Bat-Inspired Flapping Robot. *Biomimetics* **2025**, *10*, 161. [CrossRef] [PubMed]
28. Zhang, R.; Zhang, H.; Xu, L.; Xie, P.; Wu, J.; Wang, C.; Avanzini, G. Mechanism and Kinematics for Flapping Wing Micro Air Vehicles Maneuvering Based on Bilateral Wings. *Int. J. Aerosp. Eng.* **2022**, *2022*, 1–10. [CrossRef]
29. Jafferis, N.T.; Helbling, E.F.; Karpelson, M.; Wood, R.J. Untethered flight of an insect-sized flapping-wing microscale aerial vehicle. *Nature* **2019**, *570*, 491–495. [CrossRef] [PubMed]
30. Shankar, V.; Shirakawa, N.; Ishihara, D. Novel Computational Design of Polymer Micromachined Insect-Mimetic Wings for Flapping-Wing Nano Air Vehicles. *Biomimetics* **2024**, *9*, 133. [CrossRef] [PubMed]
31. Zhao, Y.; Xu, D.; Sheng, J.; Meng, Q.; Wu, D.; Wang, L.; Xiao, J.; Lv, W.; Chen, Q.; Sun, D. Biomimetic Beetle-Inspired Flapping Air Vehicle Actuated by Ionic Polymer-Metal Composite Actuator. *Appl. Bionics Biomech.* **2018**, *2018*, 3091579. [CrossRef] [PubMed]
32. Li, T.; Tang, J.; Hu, Z.; Wang, Z.; Fu, Z. Three-dimensional vibration investigation of the thin web gear pair based on Timoshenko beam. *Thin-Walled Struct.* **2023**, *184*, 110507. [CrossRef]
33. Wang, S.; Hu, B.; Wu, Z.; Zhou, Y.; Tang, J. A Comprehensive Optimization Model of Tooth Surface Parameters for the Minimization of Contact Stress of Helical Face Gears by Considering the Avoidance of Edge Contact. *Mathematics* **2022**, *10*, 3102. [CrossRef]
34. Zhan, L.; Yao, S.; Guan, C.; Zhang, D.; Wang, B.; Zhong, S. Curing simulation and experimental analysis of composite segmented tools for aerospace applications. *Polym. Compos.* **2025**, 1–15. [CrossRef]
35. Zhao, G.; Shi, J.; Xu, W.; Sun, N.; Zeng, J.; Yang, G.; Song, K.; Zheng, J. A Study on the Failure Behavior and Force Transmission of Composite Skin-Stringer Structures Under a Compressive Load. *Materials* **2025**, *18*, 1380. [CrossRef] [PubMed]
36. Wang, H.; Zhong, X.Y.; Jia, H.; Zhang, L.W.; Liu, H.S.; Sun, M.C.; Liu, T.W.; Bao, J.W.; Bai, J.B.; Ge, S.C. Study on the Transverse Properties of T800-Grade Unidirectional Carbon Fiber-Reinforced Polymers. *Materials* **2025**, *18*, 816. [CrossRef] [PubMed]
37. Szafran, K.; Jurak, M.; Mroczka, R.; Wiacek, A.E. Preparation and Surface Characterization of Chitosan-Based Coatings for PET Materials. *Molecules* **2023**, *28*, 2375. [CrossRef] [PubMed]
38. Aktas, C.; Bhethanabotla, V.; Ayyala, R.S.; Sahiner, N. Modifying wetting properties of PI Film: The impact of surface texturing and CF_4 and O_2 plasma treatment. *Appl. Surf. Sci.* **2024**, *656*, 159729. [CrossRef]
39. Zhang, C.; Zhao, M.; Wang, L.; Yu, M. Effect of atmospheric-pressure air/He plasma on the surface properties related to ink-jet printing polyester fabric. *Vacuum* **2017**, *137*, 42–48. [CrossRef]

materials

MDPI

Article

Dynamic Behaviour of Two-Layered Beam Subjected to Mechanical Load in Thermal Environment

Simona Doneva [1,2], Jerzy Warminski [3] and Emil Manoach [1,2,*]

1 Institute of Mechanics, Bulgarian Academy of Sciences, 1113 Sofia, Bulgaria; s.doneva@imbm.bas.bg
2 Centre of Excellence in Informatics and Information and Communication Technologies, 1113 Sofia, Bulgaria
3 Department of Applied Mechanics, Lublin University of Technology, 20-618 Lublin, Poland; j.warminski@pollub.pl
* Correspondence: e.manoach@imbm.bas.bg

Abstract

The research investigates a composite beam composed of two layers of different materials and thicknesses subjected to thermal and mechanical loads. Two cases of thermal loading are considered here: uniformly distributed temperature along the whole beam and linearly distributed temperature along the beam thickness. A reduced model of the problem based on the first three beam normal modes is formulated. Additionally, a simplified one-mode reduction model is developed and solved analytically by the harmonic balance method (HBM). A comparison between the results of the three-mode reduction and one-mode reduction models highlights the applicability and limitations of the latter. Differences in the resonance curves produced by these models are thoroughly examined. The correctness of the reduced models is validated through comparison with the created finite element model (FEM) of the bilayer beam. The detailed bifurcation diagrams presented for the three-degrees-of-freedom (3-DOFs) model reveal phenomena such as loss of stability, mode interaction, buckling, and existence of multiple solutions. These findings provide deeper insights into the dynamic behaviour of thin composite beams subjected to mechanical and thermal loads, considering different variations of the temperature distribution.

Keywords: nonlinear dynamics; thermoelastic; reduced model; harmonic balance method; analytical solution; 3 mode reduc-tion; bifurcation diagrams; resonance curves; AUTO

1. Introduction

In recent decades, the dynamic behaviour of thin-walled structures subjected to multi-physics fields has been a topic of increasing investigation. One explanation for this fact is the fast growth of high-technology industries. Typical examples are aerospace industries, chemical and nuclear energy industries, and micro- and nanotechnologies. Many structures applicable in these industries are subjected to the influence of temperature and electrical, magnetic, or fluid fields. The interaction between mechanical fields and concomitant temperature, electrical, or other fields could be significant for the design, proper exploitation, and safety of structures. Often, some structures are built from composite materials consisting of materials with different properties.

Thermomechanical problems are among the most commonly encountered problems in engineering practice and are extensively researched in the scientific literature. The classical books that present the basic problems of thermoelasticity are Boley and Weiner [1] and Nowacki [2].

While early work on thermoelasticity dealt primarily with static problems, the aforementioned industries are accelerating the solution of dynamic thermoelastic problems.

The evolution of dynamic thermoelastic investigations is closely related to the aerospace industry. A series of works studies the phenomena arising due to the combination of airflow, thermal, and mechanical fields acting on aerospace structures. Some of them are systematized in Torton's book [3].

Other works (referenced as [4–8]) further explore flutter, buckling, and post-buckling behaviours in panels exposed to thermal environments.

A notable characteristic of many works focused on dynamic thermoelastic problems is that the subjects of investigation are often composite structures. Most often, a homogenization procedure is applied for such structures, and the problem is transformed into one of a structure with generalized properties. For example, in [9] and [10], beams with functionally graded materials are considered, and dynamics and stability problems are solved in the case of thermal loading. Also, the dependence of material gradations on the critical buckling temperature is shown.

A large number of works investigate the nonlinear dynamic behaviour of structures subjected to mechanical and thermal loading. The interesting nonlinear phenomena arising in the process of the interaction of these two fields are a challenge to scientists. Such works include references [11–21].

In [11], the equations of motion of a simply supported rectangular plate subjected to thermal and mechanical loads are transformed to three ordinary differential equations by the Galerkin method. The authors have computed numerically various nonlinear features, including Poincaré maps, phase plots, power spectra, and bifurcation diagrams. In [12], thermoelastic vibrations of a rotating microbeam subjected to a laser pulse heat source and sinusoidal heating are considered. Mathematical modelling of the problem is developed using the Euler–Bernoulli beam theory and the generalized theory of thermal conductivity with three relaxation coefficients of time. The closed-form solutions are obtained by Laplace's transform. Laplace's transform is again applied in [13] to study the thermoelastic vibration of functionally graded nanobeams interacting with abrupt heat in nonclassical thermoelasticity with phase delays. Systematic and consistent studies of the nonlinear thermoelastic dynamic behaviour of plates are presented in [14–19]. In this series of works, the local and global dynamics of homogeneous or composite plates subjected to mechanical and thermal loads are presented. Different plate theories have been used, and coupled and uncoupled problems are considered.

The nonlinear dynamics of the coupled thermoelastic vibrations of isotropic and orthotropic plates are discussed in the work of Yen [20]. Various nonlinear dynamics features, including Poincaré maps, phase diagrams, fractal dimensions, bifurcation diagrams, power spectra, and Lyapunov exponents, are employed to describe the behaviour of these structures.

The effects of the coefficient of thermal expansion (CTE) and the thermal conductivity (TC) on the microbeam's linear free and forced vibrations are studied in [21]. The authors have used the previously developed scale-dependent linear thermoelastic model of microbeams and have applied the Galerkin method to obtain ordinary differential equations (ODEs). Then, using an analytical approach, the relationships of the bending and temperature vibrations are illustrated, and conclusions about the mechanical energy dissipation and the influence of the CTE and TC on the microbeam's behaviour are drawn.

Different aspects of dynamic thermoelastic problems of plates and beams are referenced in our previous works devoted to the same topic [22–24]. The complete model of thermoelastic vibration of plates considers transversal and longitudinal displacements, taking into account the large displacements, shear deformation, inertia terms due to cross-section rotation, and thermal and mechanical loadings. Based on the described model, the influence of temperature on the first resonance zone and the bifurcation scenario, which results in buckling and chaotic oscillations, is discussed in [22] and [23]. In [24], the reduced model is based on a three-mode reduction and considers the effect of elevated temperature on the plate's response. On the basis of an analytical model, the buckling phenomenon, post-buckling oscillations, period-doubling bifurcations, and zones of multistable solutions are demonstrated.

Bimaterial beams belong to a special class of composite structures. Usually, the two layers have quite different material properties and thicknesses. Examples of such structures can be found in many micro-electronic devices (MEMSs), energy harvesting devices, and other applications. Most studies of the thermomechanical behaviour of bimaterial beams focus on their static states. Consideration of the bending of bimaterial beams dates back to as early as 1960, as noted in the book by Boley and Weiner [1]. Later, numerous studies emerged regarding the deformation, stresses, and temperature distribution of bi-metallic beams. The static thermoelastic deformation of composite beams is studied in [25–28]. In [29], the influence of temperature on the vibrations of laminated layers made of two different materials is presented. Essentially, the object of the investigation can be considered as a functionally graded beam.

The authors of this study have applied a model of homogenization of a bilayered beam subjected to temperature changes and periodic dynamic loading [30]. A reduced model, based on a three-mode reduction, was created to study the dynamics of the beam in detail in the frequency domain. The bifurcation analysis showed the possibility of multiple solutions arising and loss of stability. The considered beam was somewhat thick, and obtaining a nonlinear response required large loads. The attempted experimental study confirmed this.

In the present work, the mathematical thermoelastic model was again formulated but for a much thinner beam, which is much more sensitive to thermal and mechanical loads. Therefore, in addition, the case of the temperature linearly distributed along the beam thickness is investigated. This makes the reduced model more complicated but allows the discovery of new phenomena that arise due to temperature influence.

The article is organized as follows: In Section 2, the mathematical model of the dynamic behaviour of a two-layered beam in a thermal environment is presented. The homogenization of the beam is explained, and the basic equations of the beam motion in terms of displacements and rotations are derived. In Section 3, the Galerkin approach is applied, and based on the first three normal modes, a reduced model is created consisting of three nonlinear ordinary differential equations. In Section 4, a particular case of a one-mode reduction model is shown, and the harmonic balance method to solve the obtained nonlinear ordinary differential equation is briefly presented. In Section 5, the FEM is applied for verification of the reduced model. In Section 6, a numerical study based on the continuation method and predictor–corrector steps is performed, and a large number of frequency response curves and bifurcation diagrams are shown. A parametric study of the influence of amplitude and frequency of loading, as well as temperature, on the beam's response is presented. In Section 7, the applicability of the one-mode reduction model is studied. In Section 8, the main conclusions are drawn.

2. Mathematical Model

Figure 1 shows a composite beam consisting of two layers of different materials (Material 1 and Material 2) with thicknesses h_1 and h_2. The length of the beam is denoted with l, the width with b, and the thickness with h, where $h = h_1 + h_2$. The coordinate along the z axis of the interface between the layers is denoted with z_m.

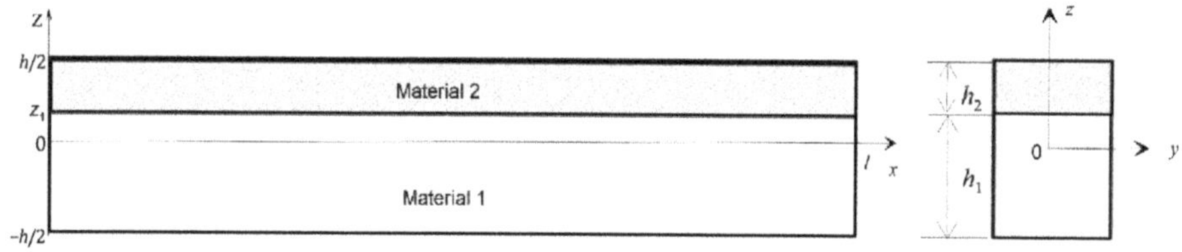

Figure 1. The geometrical scheme of the beam model.

Assuming large displacements and applying the Timoshenko beam theory, the strain and curvature-displacement relationships associated with the mid-axes can be expressed as

$$\varepsilon_x^0 = \frac{\partial u}{\partial x} + \frac{1}{2}\left(\frac{\partial w}{\partial x}\right)^2 \varepsilon_{xz}^0 = \frac{\partial w}{\partial x} - \psi\, \kappa^0 = -\frac{\partial \psi}{\partial x} \tag{1}$$

and the strain vector is expressed as follows:

$$\varepsilon = \left\{\varepsilon_x^0 + z\kappa_x^0,\ f(z)\varepsilon_{xz}^0\right\}^T \tag{2}$$

where $w(x,t)$ is the transverse displacement, $\psi(x,t)$ is the rotation angle, and $f(z)$, according to the Timoshenko beam theory, is a function describing the distribution of the shear strain along the beam thickness. The superscript ^{0} associates the strains ε_{x}, $\varepsilon_{\mathrm{xz}}$ and the curvature $\leq$ with the mid-axes.

2.1. Constitutive Equations

The relationships between the stress $\mathbf{S} = \{\sigma_x, \sigma_{xz}\}$ and strains $\varepsilon = \{\varepsilon_{\mathrm{x}},\ \varepsilon_{\mathrm{xz}}\}$, considering the temperature changes, are presented by the following well-known equations (see, for example, the basic book [31] and also the more recent monography [32]):

$$\sigma_x^{(i)} = E^{(i)}\left[\varepsilon_x - \alpha_T^{(i)}(T - T_i)\right],\ \sigma_{xz}^{(i)} = G^{(i)}\varepsilon_{xz}\ i = 1,2 \tag{3}$$

where $E^{(\mathrm{i})}$ is the Young's modulus of ith layer, $G^{(\mathrm{i})}$ is the corresponding shear modulus, $\alpha_T^{(i)}$ is the coefficient of thermal expansion, T is the current temperature, and T_i is the initial (environmental) constant temperature.

2.2. Equations of Motion

Starting from the equilibrium equations written for each layer and then following the procedure for homogenization described in [30], the following equations are obtained:

$$\frac{\partial^2 u}{\partial x^2} = G_u + G_u^T \tag{4}$$

$$\frac{\partial^2 \psi}{\partial x^2} - \frac{\overline{Gh}}{\overline{EIZ}}k^2\left(\frac{\partial w}{\partial x} - \psi\right) - c_2\frac{\partial \psi}{\partial t} - \frac{\overline{\rho I}}{\overline{EIZ}}\frac{\partial^2 \psi}{\partial t^2} = G_1^T \tag{5}$$

$$k^2\frac{\overline{Gh}}{\overline{Eh}}\left(\frac{\partial^2 w}{\partial x^2} - \frac{\partial \psi}{\partial x}\right) - c_1\frac{\partial w}{\partial x} - \frac{\overline{\rho h}}{\overline{Eh}}\frac{\partial^2 w}{\partial t^2} = -p_1(x,t) + G_2^L + G_2^T \tag{6}$$

where c_1 and c_2 denote damping coefficients which are assumed to be proportional to the mass terms in front of the inertia terms. However, the inertia term in the longitudinal direction is neglected.

In Equations (4)–(6), the following notations are introduced:

$$\overline{EI} = \frac{1}{3}\left\{E^{(1)}\left(z_1^3 + \frac{h^3}{8}\right) + E^{(2)}\left(\frac{h^3}{8} - z_1^3\right)\right\}, \tag{7}$$

$$\overline{EIZ} = \left[\overline{EI} - \frac{\left(E^{(1)} - E^{(2)}\right)^2}{\overline{Eh}} \frac{\left(h^{(1)}h^{(2)}\right)^2}{4}\right] \tag{8}$$

$$G_u = -\left(\frac{\partial w}{\partial x}\frac{\partial^2 w}{\partial x^2}\right) - \frac{\left(E^{(1)} - E^{(2)}\right)}{\overline{Eh}} \frac{h^{(1)}h^{(2)}}{2}\frac{\partial^2 \psi}{\partial x^2} \tag{9}$$

$$G_u^T = \frac{1}{\overline{\overline{Eh}}}(E^{(1)}\alpha_T^{(1)}\frac{\partial \gamma_T^{(1)}}{\partial x} + E^{(2)}\alpha_T^{(2)}\frac{\partial \gamma_T^{(2)}}{\partial x}) \tag{10}$$

$$G_1^T == -\left(E^{(1)}\alpha_T^{(1)}\frac{\partial}{\partial x}\chi_T^{(1)} + E^{(2)}\alpha_T^{(2)}\frac{\partial}{\partial x}\chi_T^{(2)}\right)\frac{1}{\overline{\overline{EIZ}}} \tag{11}$$

$$G_2^L = -\left[\left(\frac{\partial u}{\partial x} + 0.5\left(\frac{\partial w}{\partial x}\right)^2\right) + \frac{\left(E^{(1)} - E^{(2)}\right)h^{(1)}h^{(2)}}{2\overline{Eh}}\frac{\partial \psi}{\partial x}\right]\frac{\partial^2 w}{\partial x^2} \tag{12}$$

$$G_2^T = \left[E^{(1)}\alpha_T^{(1)}\gamma_T^{(1)} + E^{(2)}\alpha_T^{(2)}\gamma_T^{(2)}\right]\frac{\partial^2 w}{\partial x^2}\frac{1}{\overline{\overline{Eh}}} \tag{13}$$

$$p_1(x,t) = \frac{p(x,t)}{\overline{Eh}} \tag{14}$$

$$\begin{gathered} \gamma_T^{(1)} = \int_{-h/2}^{z_1} Tdz,\ \gamma_T^{(2)} = \int_{z_1}^{h/2} Tdz;\ \chi_T^{(1)} = \int_{-h/2}^{z_1} Tzdz,\ \chi_T^{(2)} = \int_{z_1}^{h/2} Tzdz, \\ \gamma_T^{(1)} = \int_{-h/2}^{z_1} Tdz,\ \gamma_T^{(2)} = \int_{z_1}^{h/2} Tdz\ ;\ \chi_T^{(1)} = \int_{-h/2}^{z_1} Tzdz,\ \chi_T^{(2)} = \int_{z_1}^{h/2} Tzdz, \end{gathered} \tag{15}$$

The equation describing the temperature propagation in the beam, without considering full coupling with the mechanical field, is

$$\frac{c_p^{(i)}}{\lambda_T^{(i)}}\frac{\partial T}{\partial t} = \frac{\partial^2 T}{\partial x^2} + \frac{\partial^2 T}{\partial z^2} \quad i = 1,2 \tag{16}$$

where $c_p^{(i)}$ and $\lambda_T^{(i)}$ are the heat capacity per unit volume and the thermal conductivity of the ith material.

In this work, the temperature field is established, and the temperature is uniformly distributed along the beam length. This means that the temperature does not change in time and the derivatives along the x axis are zero. Therefore,

$$G_u^T = 0 \text{ and } G_1^T = 0. \tag{17}$$

Concerning the distribution of the temperature along the beam thickness, two cases are considered: when $\Delta T = T - T_i$, the temperature is elevated/lowered along the whole beam, and in the second case, the temperature is distributed along the beam thickness following the linear law

$$T(z) = T_0 + T_1 z \tag{18}$$

The constants T_0 and T_1 are defined by the temperature boundary conditions.

The following boundary conditions regarding the temperature are defined:

$$\begin{aligned} T(-h/2) &= T_d \\ T(h/2) &= T_{up} \end{aligned} \tag{19}$$

where T_d and T_{up} denote temperature on the bottom (down) and upper beam surfaces.

In this case, the coefficients in Equation (18) are

$$T_0 = \frac{T_{up} + T_d}{2},\ T_1 = \frac{T_{up} - T_d}{h} \tag{20}$$

Considering this (see Figure 1),

$$\begin{aligned} h_2 &= h/2 - z_m \\ h_1 &= z_m + h/2 \end{aligned} \tag{21}$$

it is possible to obtain the case of linear distribution of the temperature along the beam thickness:

$$\begin{aligned} \gamma_T^{(1)} &= \int_{-h/2}^{z_1} (T_0 + T_1 z)dz = T_0 h_1 - 0.5 T_1 h_1 h_2 \\ \gamma_T^{(2)} &= \int_{z_1}^{h/2} (T_0 + T_1 z)dz = T_0 h_2 + 0.5 T_1 h_1 h_2 \end{aligned} \tag{22}$$

while the following applies in the case of uniform temperature distribution:

$$\gamma_T^{(i)} = h_i \Delta T \qquad i = 1,2 \tag{23}$$

In the further text, the notation ΔT means a difference between the beam and ambient temperatures.

3. Reduced Model

The mathematical model of the beam vibration is reduced by means of the Galerkin procedure. To transform the partial differential equations into ordinary differential equations, the first three normal modes of the free vibrations of the homogenized beam are used as basis functions.

For convenience, the following dimensionless variables are introduced:

$$\widetilde{w} = w/l,\ \widetilde{u} = u/l,\ \widetilde{\psi} \equiv \psi,\ \widetilde{T} = (T - T_0)/T_d\ \ \widetilde{x} = x/l,\ \ \widetilde{t} = tc/l,\ c^2 = \frac{\overline{EIZ}}{\overline{\rho I}} \tag{24}$$

Then, Equation (4) is transformed into

$$\frac{\partial^2 \widetilde{u}}{\partial \widetilde{x}^2} = \widetilde{G}_u \tag{25}$$

$$\frac{\partial^2 \psi}{\partial x^2} - d_2 \dot{\psi} - \alpha\beta\left(\frac{\partial \widetilde{w}}{\partial \widetilde{x}} - \psi\right) - \ddot{\psi} = 0 \tag{26}$$

$$\beta\left(\frac{\partial^2 \widetilde{w}}{\partial x^2} - \frac{\partial \psi}{\partial x}\right) - d_1 \frac{\partial \widetilde{w}}{\partial t} - s\frac{\partial^2 \widetilde{w}}{\partial t^2} = -\widetilde{p}_1(x,t) + \widetilde{G}_2^L + \widetilde{G}_2^T \tag{27}$$

where the following notations are introduced in Equations (26) and (27):

$$\beta = k^2 \frac{\overline{Gh}}{\overline{Eh}},\ \alpha = \frac{\overline{Eh}l^2}{\overline{EIZ}},\ s = \frac{\overline{\rho h}}{\overline{Eh}} \frac{\overline{EIZ}}{\overline{\rho I}} \tag{28}$$

and Equations (12)–(14) are transformed into

$$\widetilde{p}_1(x,t) = \frac{lp(x,t)}{\overline{E}h}, \tag{29}$$

$$\widetilde{G}_2^L = -\left[\left(\frac{\partial \widetilde{u}}{\partial x} + 0.5\left(\frac{\partial \widetilde{w}}{\partial x}\right)^2\right) + \frac{\left(E^{(1)} - E^{(2)}\right)h^{(1)}h^{(2)}}{2\overline{E}hl}\frac{\partial \psi}{\partial x}\right]\frac{\partial^2 \widetilde{w}}{\partial x^2} \tag{30}$$

$$\widetilde{G}_2^T = h\left[E^{(1)}\alpha_T^{(1)}\gamma_T^{(1)} + E^{(2)}\alpha_T^{(2)}\gamma_T^{(2)}\right]\frac{\partial^2 \widetilde{w}}{\partial x^2}\frac{1}{\overline{\overline{E}h}} \tag{31}$$

where d_1 and d_2 are damping coefficients. For simplicity, in the text below "tilde" is omitted. The solution of Equations (26) and (27) is sought in the following form:

$$w(x,t) = \sum_{i=1}^{3} W_i(x)q_i(t),\ \psi(x,t) = \sum_{i=1}^{3} \Psi_i(x)q_i(t) \tag{32}$$

where $W_i(x)$ and $\Psi_i(x)$ are the normal modes of the beam vibrations—the solution of Equations (26) and (27) with a zero right-hand side and satisfying the boundary conditions. Substituting Equation (32) into Equations (26) and (27), using the orthogonality condition, the following system of nonlinear ordinary differential equations is obtained:

$$\ddot{q}_n + 2\xi_n\dot{q}_n + \omega_n^2 q_n = F_n^p + F_n^L + F_n^T \tag{33}$$

In Equation (33), subscript "n" indicates the mode number, ω_n represents the natural frequencies of the linear elastic undamped bimaterial Timoshenko beam, ξn represents the modal damping coefficients, and

$$F_n^p = -\int_0^1 W_n(x)p_1(x,t)dx, \tag{34}$$

$$F_n^L = \int_0^1 \left[W_n(x)G_2^L(x,t)\right]dx \tag{35}$$

$$F_n^T = \int_0^1 \left[W_n(x)G_2^T(x,t)\right]dx \tag{36}$$

Following the procedure developed in [30], it is found that

$$G_2^L = -\int_0^1 \left(\frac{\partial w}{\partial \xi}\right)^2 d\xi \frac{\partial^2 w}{\partial x^2} \tag{37}$$

This denotes

$$I_{ij} = \int_0^1 \frac{\partial W_i}{\partial x}\frac{\partial W_j}{\partial x}dx \tag{38}$$

then, substituting Equation (32) into Equation (37), the following expression for the nonlinear force vector G_2^L is determined:

$$\begin{gathered}G_2^L=-0.5(I_{11}w_1''q_1^3+I_{22}w_2''q_2^3+I_{33}w_3''q_3^3)-\\-0.5\big[(I_{11}w_2''+2I_{12}w_1'')q_1^2q_2+(I_{22}w_1''+2I_{12}w_2'')q_1q_2^2+(I_{33}w_1''+2I_{13}w_3'')q_1q_3^2\big]\\-0.5\big[(I_{11}w_3''+2I_{13}w_1'')q_1^2q_3+(I_{22}w_3''+2I_{23}w'')q_2^2q_3+(I_{33}w_2''q_2q_3^2+2I_{23}w_3'')\big]\\-0.5(2I_{23}w_1''+2I_{12}w_3''+2I_{13}w_2'')q_1q_2q_3\end{gathered} \tag{39}$$

For term G_2^T, which accounts for the influence of temperature, the case of linear temperature distribution along the beam thickness is considered. When $T_{up} = T_d$, obviously, $T_1 = 0$, and the beam is just subjected to elevated/lowered temperature ΔT.

Then

$$G_2^T=\left[E^{(1)}\alpha_T^{(1)}\gamma_T^{(1)}+E^{(2)}\alpha_T^{(2)}\gamma_T^{(2)}\right]\frac{\partial^2 w}{\partial x^2}\frac{1}{\overline{\overline{Eh}}} \tag{40}$$

can be presented in the form

$$G_2^T=E_T\frac{\partial^2 w}{\partial x^2}\frac{1}{\overline{\overline{Eh}}} \tag{41}$$

where

$$E_T=\left[E^{(1)}\alpha_T^{(1)}\gamma_T^{(1)}+E^{(2)}\alpha_T^{(2)}\gamma_T^{(2)}\right]\frac{1}{\overline{\overline{Eh}}}=E_{T1}T_0+E_{T2}T_1 \tag{42}$$

$$E_{T1}=\frac{E^{(1)}h_1\alpha_T^{(1)}+E^{(2)}h_2\alpha_T^{(2)}}{\overline{Eh}},\ E_{T2}=\frac{h_1h_2}{h}\frac{E^{(2)}\alpha_T^{(2)}-E^{(1)}\alpha_T^{(1)}}{\overline{Eh}} \tag{43}$$

Therefore

$$\begin{gathered}F_n^T=\int_0^1\left[W_n(x)G_2^T(x,t)\right]dx=E_T\left(\int_0^1 W_n(W_1''q_1+W_2''q_2+W_3''q_3)dx\right)=\\(E_T)\times(R_{n1}q_1+R_{n2}q_2+R_{n3}q_3)\ n=1,2,3\end{gathered} \tag{44}$$

$$R_{ij}=\left(\int_0^1 W_i(W_j'')dx\right)\ i,j=1,2,3 \tag{45}$$

Introducing the notations

$$\begin{gathered}D_{1n1}=E_{T1}\int_0^1 W_nW_1''dx,\ D_{1n2}=E_{T1}\int_0^1 W_nW_2''dx,\ D_{1n3}=E_{T1}\int_0^1 W_nW_3''dx\\D_{2n1}=E_{T2}\int_0^1 W_nW_1''dx,\ D_{2n2}=E_{T2}\int_0^1 W_nW_2''dx,\ D_{2n3}=E_{T2}\int_0^1 W_nW_3''dx\end{gathered} \tag{46}$$

and taking into account Equations (34)–(46), the set of ODEs representing the three-mode reduced model of the large-amplitude thermoelastic vibrations of the bimaterial beam takes the following final form:

$$\begin{gathered}\ddot{q}_n+2\xi_n\dot{q}_n+\omega_n^2q_n+C_{(n)111}q_1^3+C_{(n)222}q_2^3+C_{(n)333}q_3^3+C_{(n)112}q_1^2q_2+C_{(n)113}q_1^2q_3+\\C_{(n)122}q_1q_2^2+C_{(n)223}q_2^2q_3+C_{(n)133}q_1q_3^2+C_{(n)233}q_2q_3^2+C_{(n)123}q_1q_2q_3-\ n=1,2,3\\(T_0D_{1n1}+T_1D_{2n1})q_1+(T_0D_{1n2}+T_1D_{212})q_2+(T_0D_{1n3}+T_1D_{2n3})q_3=-F_n\sin(\omega t)\end{gathered} \tag{47}$$

where

$$C_{(n)ijj} = 0.5\int_0^1 W_n(I_{22}W_i'' + 2I_{ij}W_j'')dx,\ C_{(n)123} = \int_0^1 W_n(I_{23}W_1'' + I_{12}W_3'' + I_{13}W_2'')dx$$
$$D_{1ni} = E_{T1}\int_0^1 W_nW_i''dx,\ D_{2ni} = E_{T2}\int_0^1 W_nW_i''dx\ i = 1,2,3;\ n = 1,2,3 \tag{48}$$

The values of the variables, involved in Equation (47) for the cases considered in the numerical examples are given in Appendix A.

4. Analytical Solutions for the One-Mode Reduction Model

In parallel with the 3-DOFs model, a one-mode reduction model is created and studied separately. The goal of this analysis is to derive an analytical solution to the equation of motion and compare the results with those based on three-mode reduction. The analytical solution allows us to obtain in a closed form the dependence of different parameters of the problem—excitation frequency, amplitude of the load, and temperature—on the system response. This is an essential advantage of such a reduction.

The 1-DOF model is obtained automatically by substituting in Equation (47) $q_2 = 0$ and $q_3 = 0$ and just taking into account the generalized coordinate q_1.

In this case, Equation (47) for $n = 1$ is transformed into the following equation:

$$\ddot{q}_1 + 2\xi_1\dot{q}_n + \omega_1^2 q_1 + C_{(1)111}q_1^3 - (T_0D_{111} + T_1D_{211})q_1 = -F_1\sin(\omega t) \tag{49}$$

The reduced nonlinear one-degree-of-freedom model with cubic nonlinearity is analytically solved using the harmonic balance method [33]. The solution to the equation is sought in the following form:

$$x = A_1(t)\sin\omega t + A_2(t)\cos\omega t \tag{50}$$

where A_1 and A_2 are unknown amplitudes assumed as slow functions of time.

Substituting Equation (50) into Equation (49), neglecting higher-order harmonics and assuming that the terms $\ddot{A}_1$ and $\ddot{A}_2$, along with terms involving amplitude derivatives of an order higher than one, are small quantities of a higher order, a set of first-order equations is obtained:

$$\begin{array}{l}(\omega_1^2 - \omega^2)A_1 - 2\xi\omega_1\omega A_2 + \frac{3}{4}\gamma(A_1^3 + A_1A_2^2) + \lambda\Delta TA_1 - 2\omega\dot{A}_2 + 2\xi\omega_1\dot{A}_1 = P \\ (\omega_1^2 - \omega^2)A_2 + 2\xi\omega_1\omega A_1 + \frac{3}{4}\gamma(A_2^3 + A_1^2A_2) + \lambda\Delta TA_2 + 2\omega\dot{A}_1 + 2\xi\omega_1\dot{A}_2 = 0\end{array} \tag{51}$$

Equation (51) comprises first-order time derivatives of the amplitudes and can be transformed into the so-called modulation equations, as they represent modulation of the amplitudes A_1 and A_2:

$$\begin{array}{l}\dot{A}_1(t) = -\frac{1}{16(\xi^2\omega_1{}^2+\omega^2)}\left(\begin{array}{l}-8P\xi\omega_1 + 8\lambda\Delta T\xi\omega_1A_1(t) + 8\xi\omega_1^3A_1(t) + 8\xi\omega_1\omega^2A_1(t) + \\ 6\gamma\xi\omega_1A_1^3(t) + 8\lambda\Delta T\omega A_2(t) + 8\omega_1^2\omega A_2(t) - 16\xi^2\omega_1^2\omega A_2(t) - \\ 8\omega^3A_2(t) + 6\gamma\omega A_1^2(t)A_2(t) + 6\gamma\xi\omega_1A_1(t)A_2^2(t) + 6\gamma\omega A_2^3(t)\end{array}\right) \\ \dot{A}_2(t) = -\frac{1}{16(\xi^2\omega_1{}^2+\omega^2)}\left(\begin{array}{l}8P\omega - 8\lambda\Delta T\omega A_1(t) - 8\omega_1^2\omega A_1(t) + 16\xi^2\omega_1^2\omega A_1(t) + 8\omega^3A_1(t) - \\ 6\gamma\omega A_1^3(t) + 8\lambda\Delta T\xi\omega_1A_2(t) + 8\xi\omega_1^3A_2(t) + 8\xi\omega_1\omega^2A_2(t) + \\ 6\gamma\xi\omega_1A_1^2(t)A_2(t) - 6\gamma\omega A_1(t)A_2^2(t) + 6\gamma\xi\omega_1A_2^3(t)\end{array}\right)\end{array} \tag{52}$$

where $\lambda = (T_0D_{111} + T_1D_{211})$, and $\gamma = C_{(1)111}$.

For a steady state, derivatives of amplitude are zero and, therefore, after some analytical computations, a cubic algebraic equation for the amplitude modulus $A = \sqrt{{A_1}^2 + {A_2}^2}$ is obtained:

$$
\begin{gathered}
-16{F_1}^2 + 9z^3\gamma^2 + z^2(24\gamma\lambda\Delta T + 24\gamma\omega_1^2 - 24\gamma\omega^2) + \\
z(16\lambda^2\Delta T^2 + 32\lambda\Delta T\omega_1^2 + 16\omega_1^2 - 32\lambda\Delta T\omega^2 - 32\omega_1^2\omega^2 + 64\xi^2\omega_1^2\omega^2 + 16\omega^4) = 0
\end{gathered} \tag{53}
$$

where $z = A^2$.

The modulation Equation (52) can be written in a short form as the following:

$$
\begin{aligned}
\frac{dA_1}{dt} &= f_1(A_1, A_2) \\
\frac{dA_2}{dt} &= f_2(A_1, A_2)
\end{aligned} \tag{54}
$$

Stability analysis is based on the Jacobian matrix and involves evaluating how a system responds to small disturbances near an equilibrium (fixed) point. The type and stability of each singular point depends on the eigenvalues (roots) of the Jacobian matrix:

$$
J = \begin{pmatrix} \frac{\partial f_1}{\partial A_1} & \frac{\partial f_1}{\partial A_2} \\ \frac{\partial f_2}{\partial A_1} & \frac{\partial f_2}{\partial A_2} \end{pmatrix} \tag{55}
$$

Eigenvalues may be real or complex numbers, leading to different dynamic behaviours. The solution is unstable if at least one real part of the roots is positive.

The details of the stability analysis are omitted here for the sake of simplicity.

This algebraic equation connects the excitation amplitude and frequency as well as the temperature loading and allows the resonance curves to be obtained and other parametric studies of the bimaterial beam to be performed. The analytical solution obtained by the HBM will be compared with the solution for the more advanced three-mode model obtained directly from ODEs by the continuation method [34]. This comparison gives information about the applicability of one-mode reduction and demonstrates the influence of the higher-order modes on the beam dynamics, stability of the solutions, and temperature influence.

5. Finite Element Model Validation

The developed model for the dynamic behaviour of the thermoelastic bimaterial beam is validated through finite element modelling. A three-dimensional model of the bimaterial beam is created using the commercial software ANSYS 2021/R2 (Canonsburg, PA, USA) [35]. The three-dimensional finite element SOLID 226 is used for discretization of the beam. It is applicable to a lot of coupled-fields analyses including Structural–Thermal. This element has twenty nodes with up to six degrees of freedom per node, which allows computations with a high accuracy.

First, a mesh convergence study focusing on the natural frequencies, critical buckling temperature, and maximum displacement is performed. Three different levels of mesh density are created: coarse mesh (6000 elements and 8844 nodes), medium mesh (32,000 elements and 157,889 nodes), and fine mesh (256,000 elements and 295,569 nodes). The results obtained by these three different models show that the quantities of interest converge with mesh refinement, with relative changes being much less than 1% between the medium and fine meshes—see Table 1.

Table 1. Natural frequencies obtained with different meshes.

Number	Mesh with 6000 Elements, Hz	Mesh with 32,000 Elements, Hz	Mesh with 256,000 Elements, Hz
1	121.78	117.39	117.22
2	335.40	323.30	322.87
3	656.84	633.20	632.38

Based on the obtained results, we verified that the medium-density mesh used in the main simulations provides a balance between accuracy and computational cost and is fine enough to capture the response of the structure. The finite element discretization of the beam is illustrated in Figure 2.

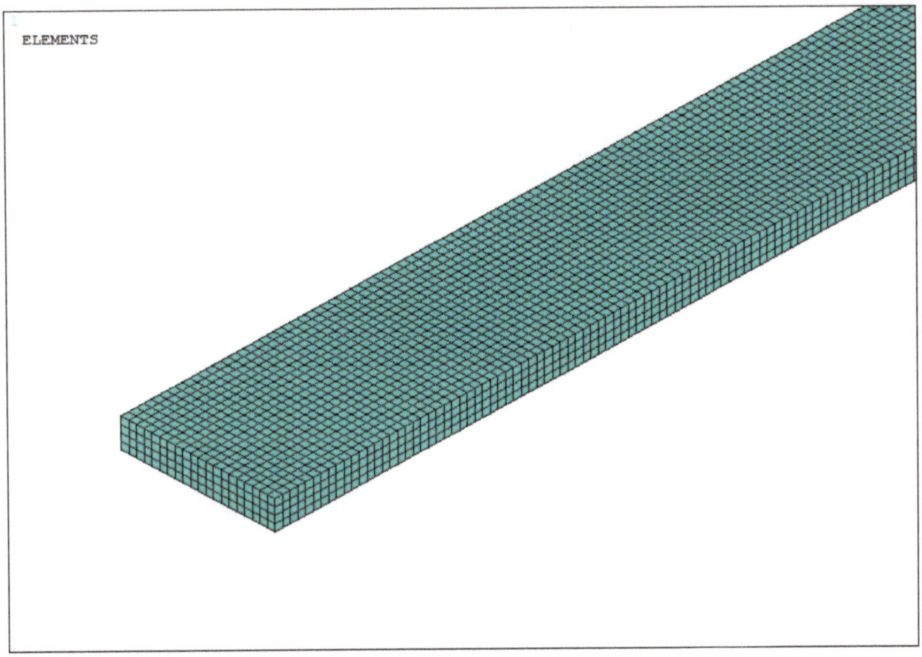

Figure 2. Discretization of the beam model.

The analysed beam is composed of two layers: the bottom layer is made of aluminium alloy Al 1050 (Material 1), and the upper layer is made of copper C 12,500 (Material 2). The dimensions of the beam shown in Figure 1 are as follows:

$$l = 0.4 \text{ m}, h = 0.004 \text{ m}, h_1 = 0.003 \text{ m}, h_2 = 0.001 \text{ m}, b = 0.02 \text{ m}$$

with the following material parameters of each layer:

$$E^{(1)} = 7.0 \times 10^{10} \text{ N/m}^2, \rho^{(1)} = 2778 \text{ kg/m}^3, \nu_1 = 0.34, \alpha_T{}^{(1)} = 23.9 \times 10^{-6} \text{ 1/K}$$

$$E^{(2)} = 12.8 \times 10^{10} \text{ N/m}^2, \rho^{(2)} = 8940 \text{ kg/m}^3, \nu_2 = 0.34, \alpha_T{}^{(2)} = 16.7 \times 10^{-6} \text{ 1/K}$$

To express the nonlinear phenomena more clearly, clamped–clamped boundary conditions are assumed in the analysis.

The thickness of the beam is discretized into four layers: one layer for copper and three layers for aluminium.

The initial task is to perform a modal analysis to determine the natural frequencies and the corresponding mode shapes. The natural frequencies calculated by the finite element model are compared with those obtained analytically for the reduced model, and the results are presented in Table 2.

Table 2. The natural frequencies of the beam according to the finite element model and the Timoshenko beam theory for the homogenized beam.

No	Reduced Model Hz	Finite Element Model Hz	Difference %
1	108.77	117.39	7.3
2	299.53	323.30	7.35
3	586.42	633.20	7.38

A difference of about 7% was also recorded in a previous paper using homogenization of the beam and the reduced model [30].

The second task is to compare the forced response of the beam subjected to harmonic and thermal loads.

The Finite Element Method (FEM) is a powerful tool for analysing complex structures and solving nonlinear differential equations, but it requires significant computational resources and time to obtain accurate results. The limitations in terms of computational power and time can restrict the number of cases that can be studied using the FEM. However, the primary objective of FEM computations is to highlight the main trends of the analysed processes and validate the mathematical models used.

The comparison of the time history diagrams for the bimaterial beam subjected to harmonic loads with different amplitudes and temperatures, obtained using the reduced model of the beam and the finite element model, is presented in Figure 3a–c.

The presented modal analysis and computed dynamic time responses for the finite element model and the reduced three-mode model are in good agreement. The observed deviation results from different assumptions in both models. However, the differences are acceptable, and the correctness of the theoretical results is confirmed. A more advanced bifurcation analysis and study of several interesting phenomena will be performed based on the reduced models.

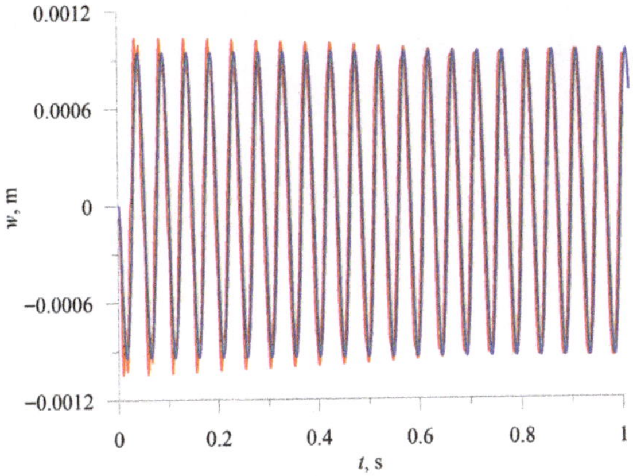

(a)

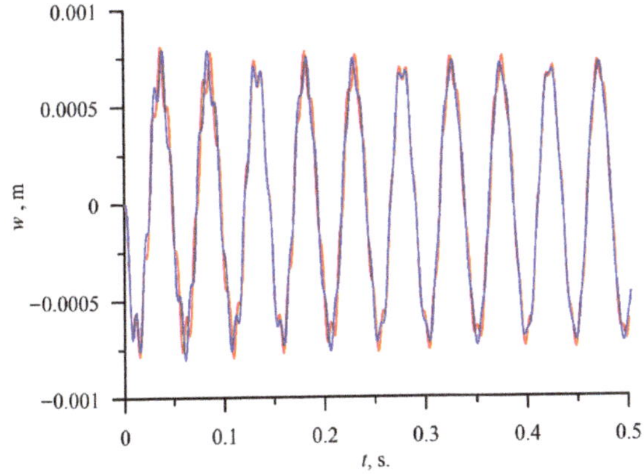

(b)

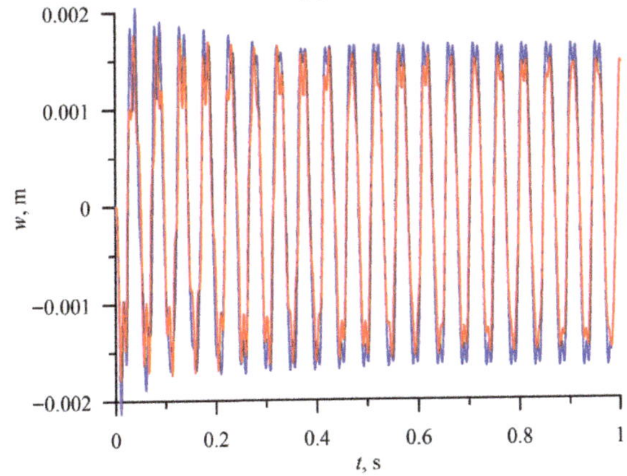

(c)

Figure 3. (**a**) Time history diagram of the response of the beam centre—red colour, ANSYS; blue colour, reduced model (3-mode reduction). T = 5, P = 5000 N/m^2, and ω = 130 rad/s. (**b**) Time history diagram of the response of the beam centre—red colour, ANSYS; blue colour, reduced model (3-mode reduction). T = 10, P = 500 N/m^2, and ω = 130 rad/s. (**c**) Time history diagram of the response of the bimaterial beam subjected to harmonic loading. P = 5000, ω = 130 rad/s, and T = 10. Comparison between the reduced model and finite element model. Blue colour—reduced model; red colour—ANSYS.

6. Nonlinear Dynamics Based on the Three-Mode Reduction Model

The equations of motion (31) of the reduced model of the thermoelastic beam's dynamics are solved numerically, and the continuation technique with predictor–corrector steps is used to obtain bifurcation curves [34]. In this case, the stability of the periodic solutions is determined by Floquet multipliers computed for the three ordinary differential equations of motion. The analysis starts with computation of the resonance curves, neglecting influence of the thermal loading, assuming $T_{up} = 0$ and $T_d = 0$. The curves are plotted for the case when the first mode is excited directly; thus, periodic force occurs on the right side of the first equation $F_1 = f_1 \sin\omega t$, while $F_2 = 0$ and $F_3 = 0$. This is the most important case because the first mode plays a crucial role in the dynamics. As the system is nonlinear, the other modes are coupled, and therefore, they are indirectly activated as well. The three levels of excitation are selected to observe the beam's response for the first q1 (Figure 4a), second q2 (Figure 4b), and third q3 (Figure 4c) generalized coordinates. For the very small level of excitation amplitude (black curve), we observe almost a linear resonance curve for q1 with negligible involvement of q2 and q3 coordinates.

As excitation increases, the curves corresponding to q_1 bend towards higher frequencies (red and green colour) and coordinates q_2 and q_3 are much more involved, however, with much larger participation of coordinate q3 than q2 (note that in Figure 4b, the scale is of 1×10^{-8} order).

Thus, considering that for moderate vibrations the modal interactions are rather negligible, the results based on the one-mode reduction model can accurately capture the real system response, as presented in Section 4. Then, the solution can be determined analytically with very good accuracy.

More complicated dynamics occur for large values of excitation, as presented in Figure 5. The instability of the resonance curves occurs for larger oscillation amplitudes, located close to the peak of the black resonance curve ($f_1 = 0.2 \times 10^{-5}$) and in the middle of the red curve ($f_1 = 0.1 \times 10^{-4}$), as presented in detail in the zoomed-in view in Figure 5b. Consequently, the instability is present for coordinate q_2 in Figure 5c and coordinate q_3 in Figure 5d.

The following question arises: what happens to the system if it works in the instability zone? In Figure 6, the solutions are determined around the frequency corresponding to bifurcation points BP_1 and BP_2. Starting from the bifurcation point BP_1, an additional stable solution arises (red line), and this is an effect of a strong involvement of coordinate q_2. From the bifurcation point BP2, only unstable solutions arise, plotted by the blue dashed line. This means that for large vibrations, bifurcation points on the resonance curve occur, and from them, new solutions arise. The second vibration mode q_2 is strongly activated. Out of this zone, the first coordinate q_1 interacts mainly with the third mode q_3, while the second mode remains at a very small level. The phenomenon of stability loss can be observed for the model with three degrees of freedom, and it is not observed for the one-mode reduction (Section 4).

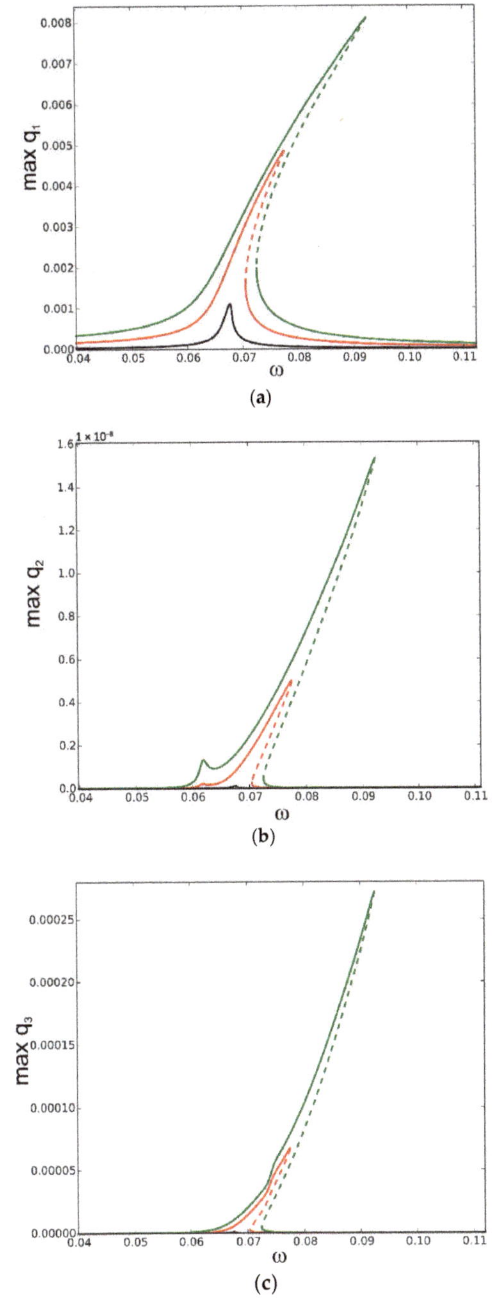

Figure 4. Resonance curves around the first natural frequency for large mechanical loading: $f_1 = 0.1 \times 10^{-6}$—black; $f_1 = 0.5 \times 10^{-6}$—red; $f_1 = 0.1 \times 10^{-5}$—green; $\Delta T = 0$; (**a**) coordinate q_1, (**b**) coordinate q_2, and (**c**) coordinate q_3.

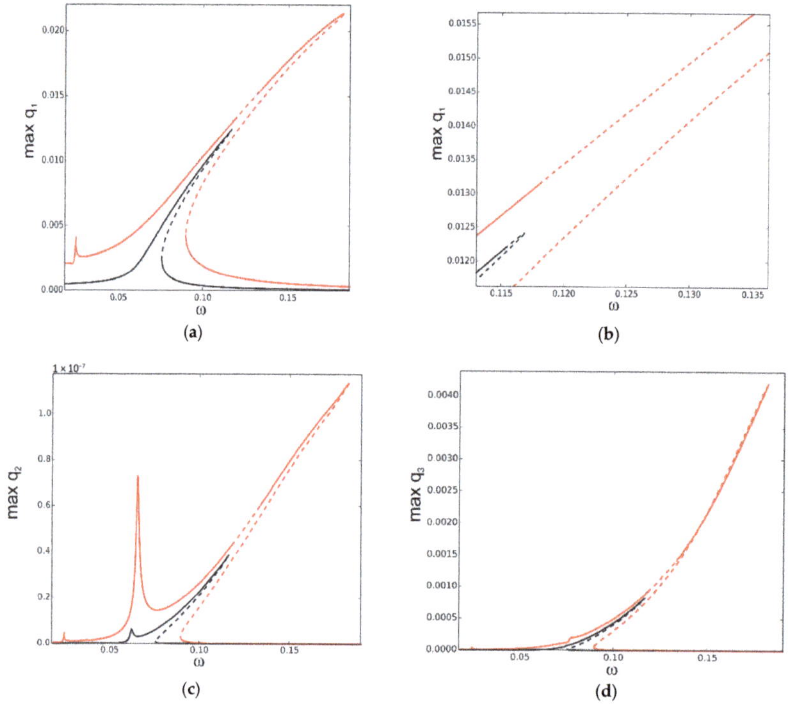

Figure 5. (**a**,**b**) Resonance curves around the first natural frequency for large mechanical loading: $f_1 = 0.2 \times 10^{-5}$—black; $f_1 = 0.1 \times 10^{-4}$—red; $\Delta T = 0$; (**a**) coordinate q_1 and (**b**) zoomed-in view of coordinate q_1 in vicinity of unstable zone. (**c**,**d**) Resonance curves around the first natural frequency for large mechanical loading: $f_1 = 0.2 \times 10^{-5}$—black; $f_1 = 0.1 \times 10^{-4}$—red; $\Delta T = 0$; (**c**) coordinate q_2 and (**d**) coordinate q_3.

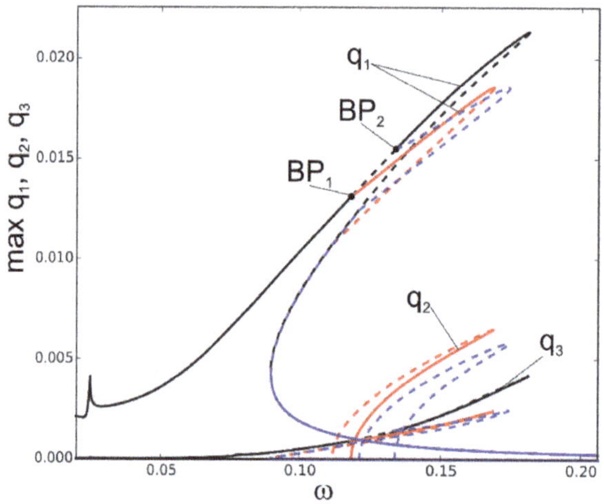

Figure 6. Resonance curves around the first natural frequency for large mechanical loading: $f_1 = 0.1 \times 10^{-4}$, $f_2 = 0$, $f_3 = 0$, and $\Delta T = 0$; coordinates q_1, q_2, and q_3 with different branches indicated by black, red, and blue colours and bifurcation points BP_1 and BP_2.

The temperature influence in the vicinity of the first natural frequency is at first demonstrated for temperature uniformly distributed through the beam volume and for $f_1 = 0.1 \times 10^{-5}$, i.e., for relatively large vibrations but without the stability loss discussed above. In Figure 7, the black curve represents the reference resonance curve for $\Delta T = 0$. The "hot" colour curves correspond to elevated temperatures $\Delta T = 5$ (orange), $\Delta T = 10$ (pink), $\Delta T = 15$ (red), and $\Delta T = 20$ (brown), and "cold" colours correspond to decreased temperatures $\Delta T = -10$ (green) and $\Delta T = -20$ (blue). A decreased temperature below the reference $\Delta T = 0$ leads to a shift of the resonance curves towards higher frequencies and to amplitude reduction. An elevated temperature up to $\Delta T = 15$ increases the amplitudes and shifts the resonance curves towards lower frequencies. At $\Delta T = 15$, the beam buckles, and a new branch arises from the bifurcation point. A further increase in the temperature up to $\Delta T = 20$ "destroys" the classical resonance curve, and just post-buckling oscillations exist (brown colour) with the partially stable branch and with the unstable part for lower frequencies. The resonance curves are also shifted correspondingly for q_2 (Figure 7b) and q_3 (Figure 7c). Their involvement in the response is more evident when the temperature is increased.

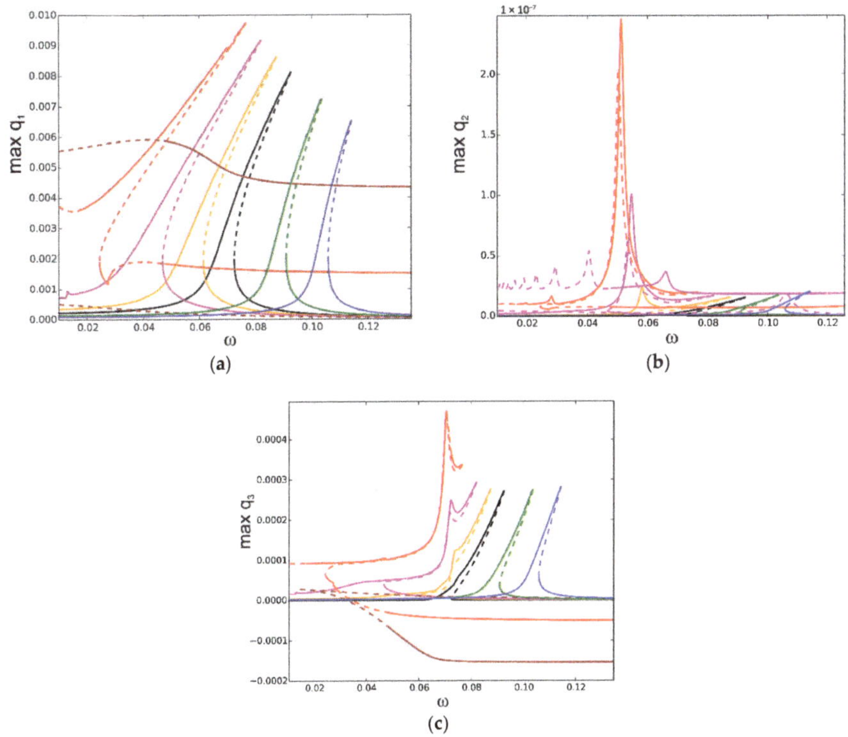

Figure 7. Resonance curves around the first natural frequency for mechanical loading $f_1 = 0.1\times 10^{-5}$ and different values of elevated temperature: $\Delta T = 0$—black; $\Delta T = 5$—orange; $\Delta T = 10$—pink; $\Delta T = 15$—red; $\Delta T = 20$—brown; $\Delta T = -10$—green; $\Delta T = -20$—blue; (**a**) coordinate q_1, (**b**) coordinate q_2, and (**c**) coordinate q_3.

The influence of the temperature is also tested for twice larger mechanical loading for $f_1 = 0.2 \times 10^{-5}$. The resonance curves are plotted for selected values of temperatures, again in "hot" colours for positive and "cold" colours for negative temperatures (Figure 8). For temperatures $\Delta T = -10$, $\Delta T = -20$, and $\Delta T = -30$, the resonance curves are shifted towards higher frequencies, presenting hardening and smaller values of amplitudes (ΔT =−10—green; ΔT = −20—blue; ΔT = −30—navy blue). In the case of elevated temperature, the response becomes more complex. For ΔT = 10, the curve is shifted towards lower frequencies (orange curve) with an additional secondary resonance demonstrated by a peak in the resonance curve. A further temperature increase up to ΔT = 15 changes the curve essentially (red colour). First of all, two instability zones on the upper branch of the curve arise, at the beginning of the curve and close to the peak. The additional phenomenon is a bifurcation of the solution with two stable and one unstable branches, which represents buckling of the beam in a positive or negative direction. The solutions for the post-buckled state are fully stable in contrast to that presented in Figure 7a for ΔT = 15 where the post-buckling instability zone exists.

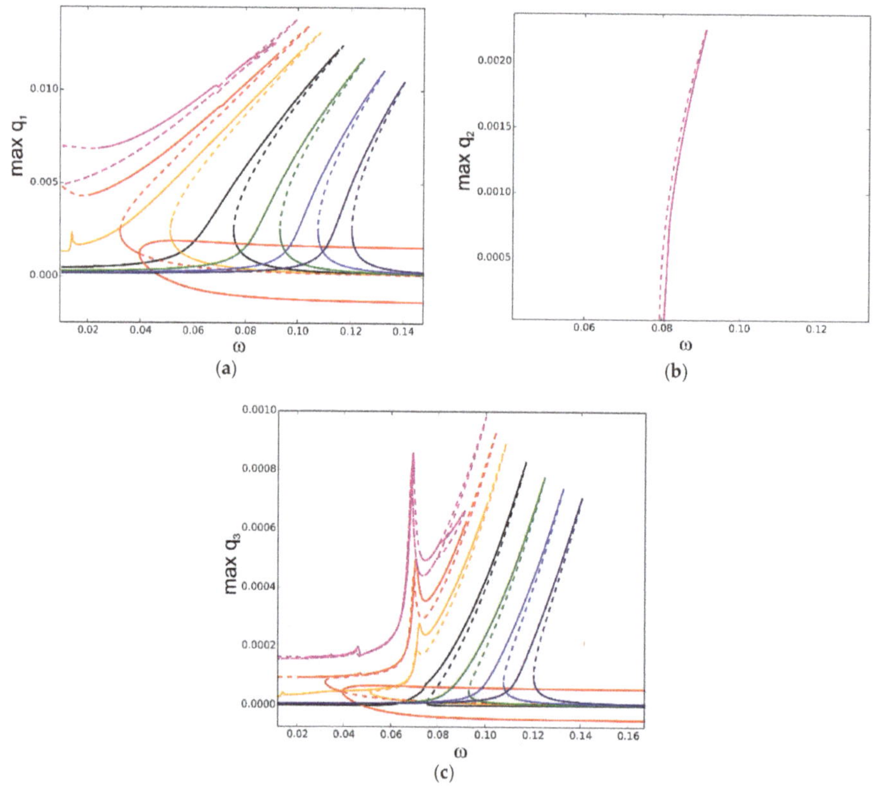

Figure 8. Resonance curves around the first natural frequency for mechanical loading $f_1 = 0.2 \times 10^{-5}$ and different values of elevated temperature: ΔT = 0—black; ΔT =10—orange; ΔT =15—red; ΔT =20—magenta; ΔT =−10—green; ΔT =−20—blue; ΔT =−30—navy blue; (**a**) coordinate q_1, (**b**) coordinate q_2, and (**c**) coordinate q_3.

For ΔT = 20, the classical shape of the resonance curve is "destroyed" (magenta curve), and close to the top of the upper branch, an additional stable solution arises. This occurs due to the strong interaction with the second mode, as presented in Figure 8b. Values of q_2 corresponding to other resonance curves are very small, and therefore, they are invisible in Figure 8b. The involvement of the third mode is presented in Figure 8c, and it remains in the same order for all discussed cases.

Analysing the above results, presented in Figures 7 and 8, we may conclude that the resonance characteristics are changed qualitatively if the temperature is elevated above about $\Delta T = 15$. To check how the response changes while the temperature is varied, the bifurcation diagrams against ΔT are computed.

The bifurcation diagram of the beam response is obtained for a mechanical excitation level of $f_1 = 0.1 \times 10^{-5}$ and for five different frequencies selected around the resonance zone. The shapes of the curves obtained against elevated temperature ΔT resemble the frequency response curves: $\omega = 0.05$—black; $\omega = 0.06$—red; $\omega = 0.07$—green; $\omega = 0.08$—blue; and $\omega = 0.10$—orange (Figure 9a). It is interesting that for all curves the bifurcation point (BP) occurs for the same temperature value, about $\Delta T = 14.35$. The involvement of the coordinate q_2 (Figure 9b) is negligible, while q_3 is one order smaller than q_1 (Figure 9c).

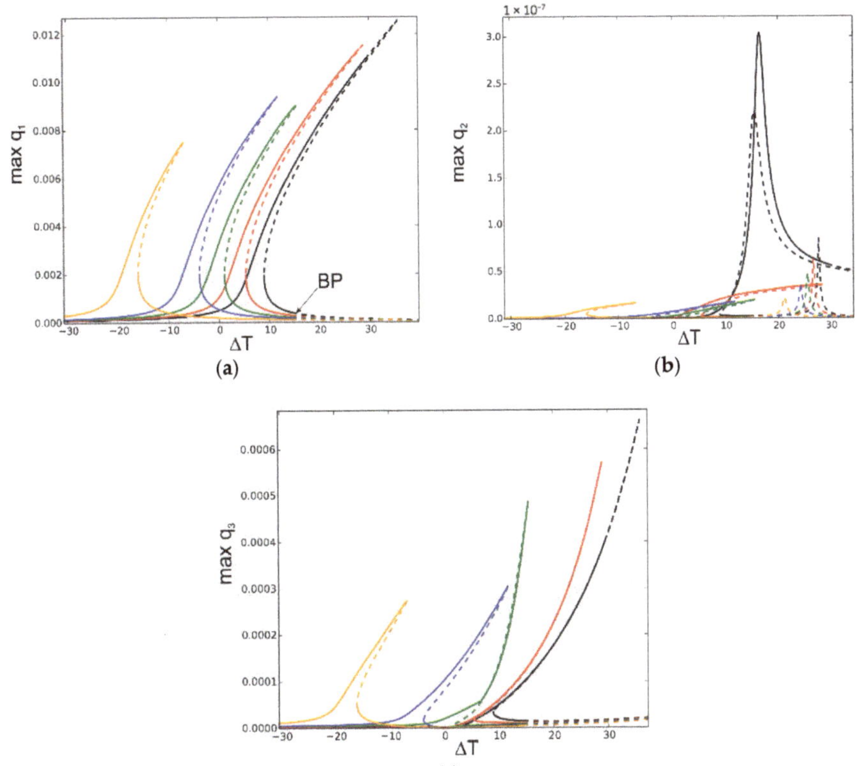

Figure 9. Bifurcation diagram of beam response against elevated temperature ΔT for fixed mechanical loading $f_1 = 0.1 \times 10^{-5}$ and $\omega = 0.07$; BP (bifurcation point) at $\Delta T = 14.35$. (**a**) Coordinate q_1, (**b**) coordinate q_2, and (**c**) coordinate q_3.

Detailed bifurcation analysis is conducted for fixed $f_1 = 0.1 \times 10^{-5}$ and $\omega = 0.07$ (Figure 10). Different branches of the solutions are indicated by dedicated colours. Starting from negative values of ΔT, we follow the black curve with the periodic response corresponding to excitation frequency. For negative temperature coordinates, q_1, q_2, and q_3 become very small values (Figure 10c,d).

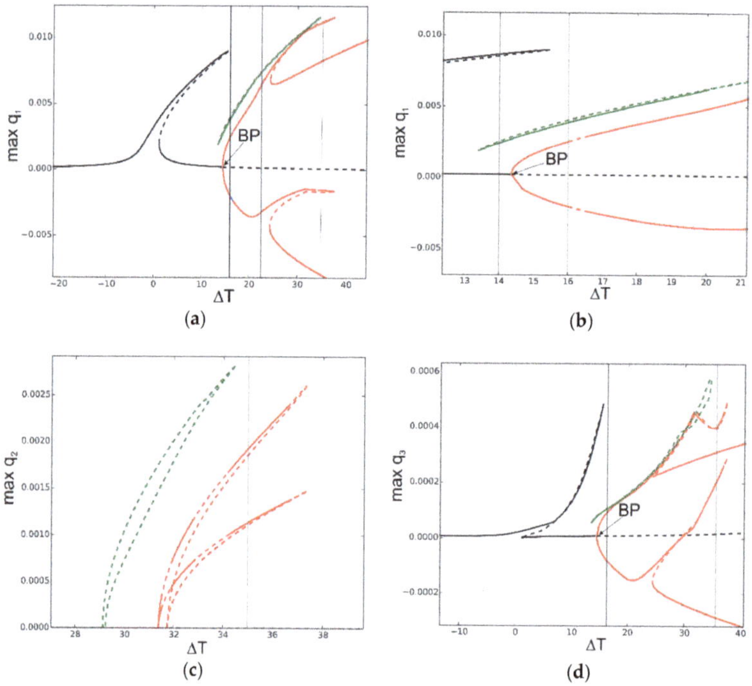

Figure 10. Bifurcation diagram of beam response against elevated temperature ΔT for fixed mechanical loading $f_1 = 0.1 \times 10^{-5}$ and $\omega = 0.07$; BP (bifurcation point) at $\Delta T = 14.35$; post-buckling oscillations—red curves; isolated solution—green colour; (**a**) coordinate q_1, (**b**) zoom of coordinate q_1 in the vicinity of BP point, (**c**) coordinate q_2, and (**d**) coordinate q_3.

Approaching $\Delta T = 0$ and going forward, the beam response increases up to the maximal value at the turning point $\Delta T = 15.43$. The bifurcation curve resembles a stiffening resonance curve with three existing solutions.

Moving forward, the periodic solution bifurcates at point BP ($\Delta T = 14.35$), which is visible for the first and third vibration modes in Figure 10a,b,d. After buckling, two branches arise (red curves). They are asymmetric, which is more visible for larger values of ΔT. Around the elevated temperature $\Delta T = 32$, the second mode is strongly involved in the beam dynamics, as presented in Figure 10c.

Apart from the periodic solutions obtained by the continuation method, the isolated solutions shown in Figure 10a by the green "isola" above the red branch are also obtained. The isolated solutions are partially stable and partially unstable (solid or dashed green line).

To demonstrate beam dynamics and coexisting solutions, the time histories are plotted for the selected temperature values: $\Delta T = 14$, $\Delta T = 16$, $\Delta T = 22$, and $\Delta T = 35$. The values of selected ΔT are indicated by thin vertical lines on the bifurcation diagram in Figure 10. The time histories are shown in Figure 11a–d. For $\Delta T = 14$ (Figure 11a), we obtain two periodic solutions corresponding to the upper and lower branches in the bifurcation diagram (black curve). However, there is a third stable solution (green line) with a period that is about tripled. This solution corresponds to the green "isola" on the bifurcation diagram. For $\Delta T = 16$ (Figure 11b), after the bifurcation point, post-buckling oscillations of the beam with a positive (red) or negative (blue) offset take place. But again, large-amplitude oscillations (green colour) exist with a tripled period, and they correspond to oscillations between both buckled states. For the higher temperature $\Delta T = 22$, just two small-amplitude vibrations around the buckled states exist, plotted by the red and blue time histories in Figure 11c. For this temperature, the solutions of the "isola" are unstable; therefore, the large-amplitude oscillations presented by a green line in Figure 11b do not exist anymore. For the higher temperature $\Delta T = 35$, four different post-buckling vibrations are possible (Figure 11d). This is a result of the third and the second mode activation, which is visible in Figure 10c,d. It means that for $\Delta T = 35$ we obtain large- and small-amplitude oscillations around the buckled states with a positive (red) and negative (blue) offset.

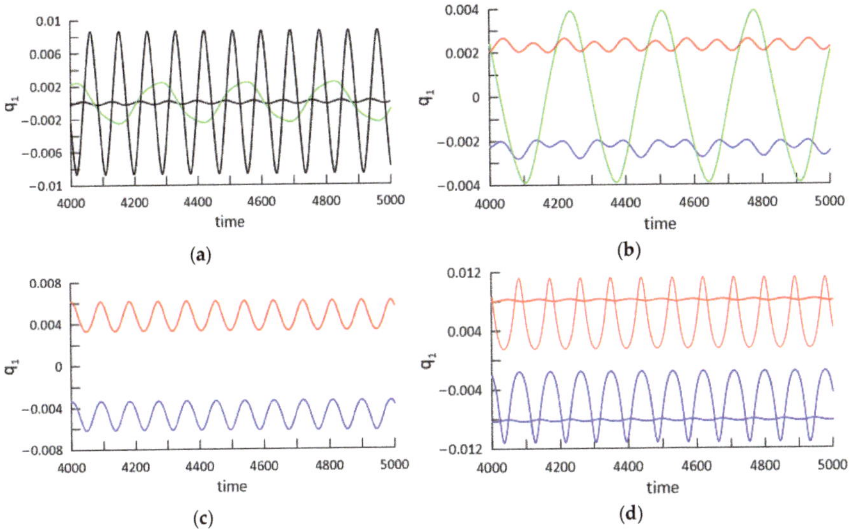

Figure 11. Time histories of beam response for selected values of elevated temperature ΔT and mechanical loading: $f_1 = 0.1 \times 10^{-5}$ and $\omega = 0.07$; (**a**) $\Delta T = 14$, (**b**) $\Delta T = 16$, (**c**) $\Delta T = 22$, and (**d**) $\Delta T = 35$.

The proposed model enables the study of the effect of heat transfer linearly distributed through the beam thickness, assuming different temperatures of the upper (T_{up}) and lower (T_d) beam surfaces. The resonance curves for different temperature gradients are shown in Figure 12. The reference resonance curve for $T_{up} = 0$ and $T_d = 0$ is plotted in black. For the temperature gradient $T_{up} = 10$ and $T_d = 0$, the resonance curve (red colour) is shifted to lower frequencies; for the opposite temperature gradient, $T_{up} = 0$ and $T_d = 10$, we observe the same trend, but the shift is smaller (orange curve). This phenomenon comes from the asymmetric geometric and material properties of the beam. In the case of negative temperature, for $T_{up} = -10$ and $T_d = 0$, the resonance curve is shifted towards higher frequencies (green curve), and for $T_{up} = 0$ and $T_d = -10$, again, the shift is smaller (blue curve).

Similar analysis is repeated for a larger temperature gradient by assuming temperatures on the upper and bottom surfaces with opposite signs. The resonance curves in Figure 13 are computed for $T_{up} = 15$ and $T_d = -15$ (black), $T_{up} = 30$ and $T_d = -30$ (red), and $T_{up} = 50$ and $T_d = -50$ (green). Comparing with the reference curve $T_{up} = 0$ and $T_d = 0$ plotted in grey, in all cases, the curves are shifted towards lower frequencies when the temperature gradient is increased.

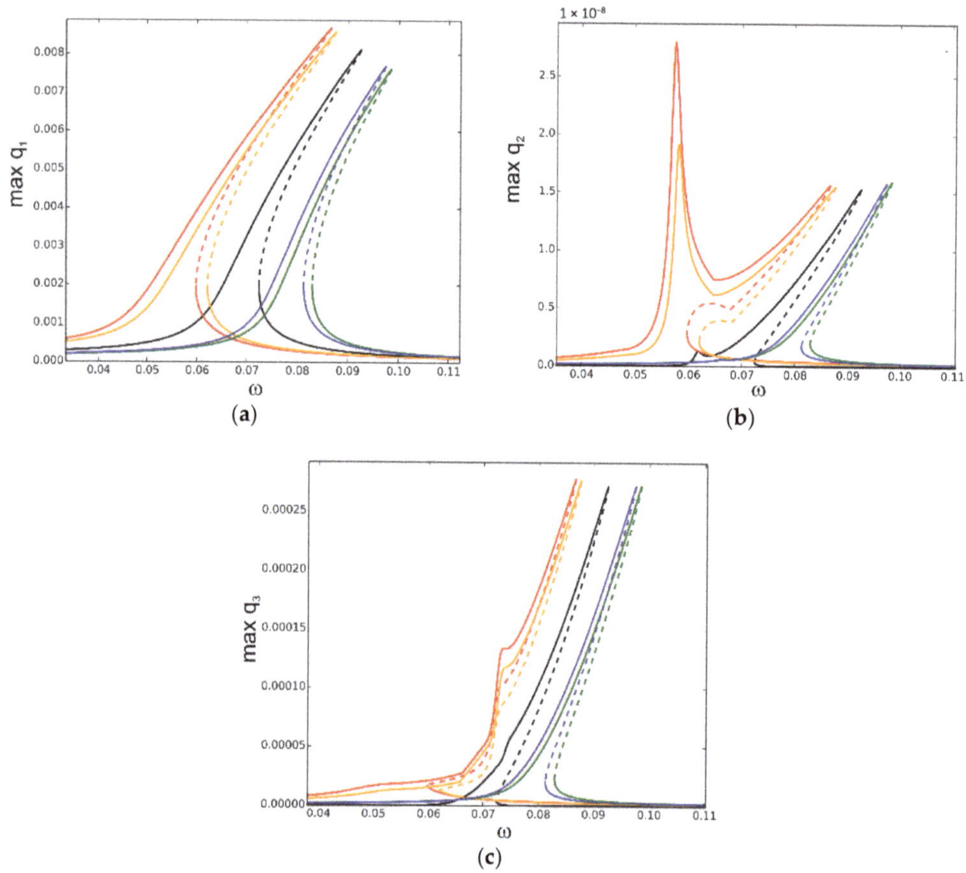

Figure 12. Resonance curves around the first natural frequency for amplitude of excitation f1 = 0.1×10^{-5} and different values of distributed temperature: T_{up} = 10, T_d = 0—red; T_{up} = 0, T_d = 10—orange; T_{up} = −10, T_d = 0—green; T_{up} = 0, T_d = −10—blue; reference curve: T_{up} = 0, T_d = 0—black; (**a**) coordinate q_1, (**b**) coordinate q_2, and (**c**) coordinate q_3.

It should be noted that for the above studied temperature gradients the beam's buckling does not occur. Even for the large temperature difference T_{up} = 50 and T_d = −50, the curve is just slightly more shifted. This is a result of two opposite effects, positive temperature on the upper part of the beam and negative on the lower part, which compensate for each other. For the elevated temperature, when the beam is uniformly heated, the buckling occurs at ΔT = 14.35.

To observe a difference in the bifurcation scenario, we select T_{up} as a bifurcation parameter while the temperature of the bottom surface is fixed, T_d = 0 (Figure 14). Then, the T_{up} temperature is varied from about T_{up} = −25 up to about T_{up} = 82. We can distinguish pre-buckling solutions (black), post-buckling (red), and isolated solutions (green). The bifurcation diagram against elevated temperature ΔT plotted in Figure 10 is now repeated as a reference in Figure 14 as a thin grey line. The bifurcation diagram against T_{up} shows a much stronger stiffening nature, and the bifurcation point of buckling takes a larger value T_{up} = 27.44 (red) compared to the BP indicated in grey. Furthermore, the post-bucking solutions and isolated green solutions are shifted towards higher temperatures.

Nevertheless, the solutions for linearly distributed temperature demonstrate qualitative similarities with the elevated temperature of the beam.

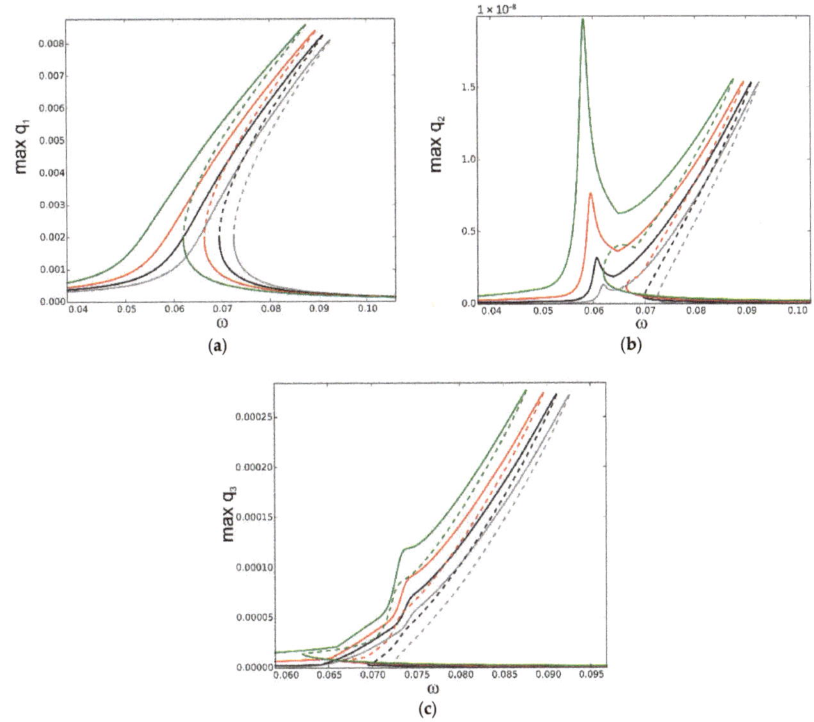

Figure 13. Resonance curves around the first natural frequency for mechanical loading $f_1 = 0.1 \times 10^{-5}$ and different gradients of linear temperature distribution: $T_{up} = 15$, $T_d = -15$—black; $T_{up} = -30$, $T_d = -30$—red; $T_{up} = 50$, $T_d = -50$—green; reference resonance curve for $T_{up} = 0$, $T_d = 0$—grey; (**a**) coordinate q_1, (**b**) coordinate q_2, and (**c**) coordinate q_3.

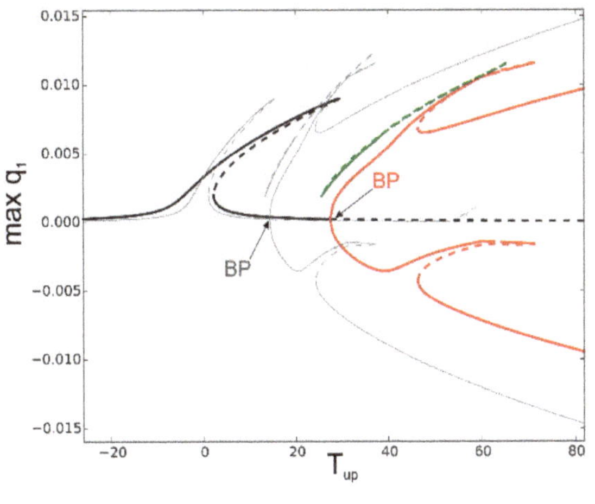

Figure 14. Bifurcation diagram of coordinate q_1 against temperature of the upper beam surface T_{up} for fixed temperature of bottom surface $T_d = 0$: reference response against elevated beam temperature ΔT—grey curve; fixed mechanical loading, $f_1 = 0.1 \times 10^{-5}$ and $\omega = 0.07$.

In Figure 15, time history diagrams for three different cases of linear distribution of the temperature along the beam thickness are shown. In all cases, the temperature of the down surface of the beam is 10, and the excitation frequency is 0.07. For a small change in temperature when $T_{up} = 15$ (black curve), the amplitude of vibrations is small. For $T_{up} = 26$ (blue curve), the temperature difference leads to additional bending, and the amplitudes of vibrations become quite large. For a bigger temperature difference, when $T_{up} = 35$ (red curve), the beam buckles and continues to vibrate around the new equilibrium state.

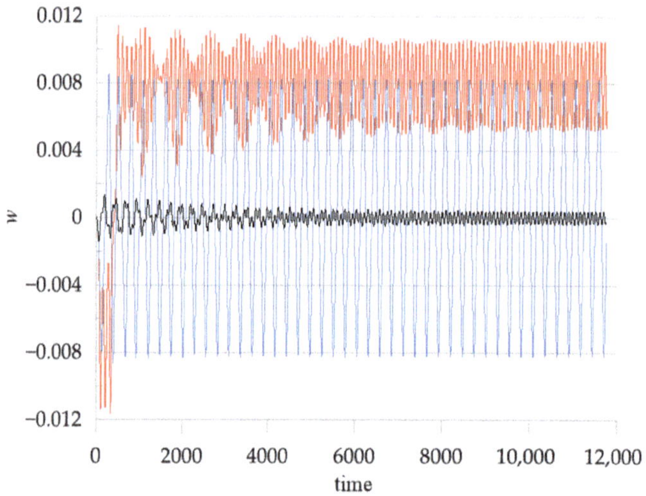

Figure 15. Time history diagrams of the response of the beam centre. Black line: $T_{up} = 15$, $T_d = 10$; blue line: $T_{up} = 26$, $T_d = 10$; red line: $T_{up} = 35$, $T_d = 10$. $f_1 = 0.1 \times 10^{-5}$ and $\omega = 0.07$.

The deflections along the beam length at a selected time (time = 6554) are shown in Figure 16. Here, the big difference in the amplitudes of the beam with just small temperature differences and the ones with bigger differences is clearly visible.

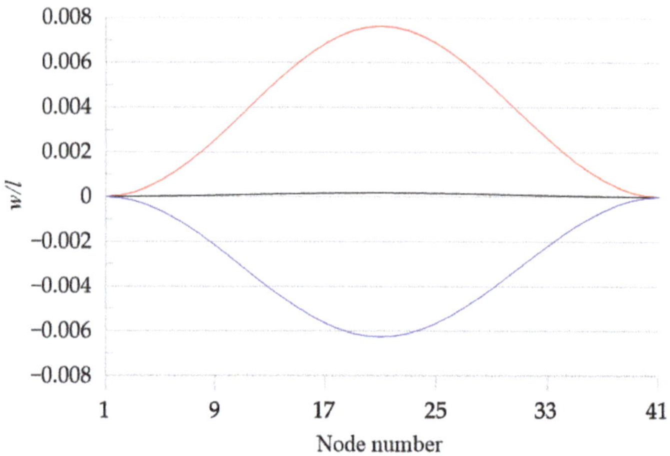

Figure 16. Displacement along the beam length for a fixed moment in time t = 6554. Black line: $T_{up} = 15$, $T_d = 10$; blue line: $T_{up} = 26$, $T_d = 10$; red line: $T_{up} = 35$, $T_d = 10$. $f_1 = 0.1 \times 10^{-5}$ and $\omega = 0.07$.

The results of the fast Fourier transforms of the time series plotted in Figure 15 are shown in Figure 17, maintaining the same colour set. The influence of the first natural frequency on the response is clearly seen in all three cases. The excitation frequency $\omega = 0.07$ is also clearly visible in the figure. It is an interesting fact that in the case of $T_{up} = 26$, for $\omega = 0.023$, a large peak in the spectra is seen with a frequency about three times smaller than the excitation frequency. This oscillation represents beam motion between two buckled states. Close to this value of frequency, there is a small peak in the curve obtained for $T_{up} = 15$. For $T_{up} = 35$, some small additional peaks appear for lower and higher frequencies, demonstrating nonlinear dynamics and the possible existence of secondary sub- or super- harmonic resonances.

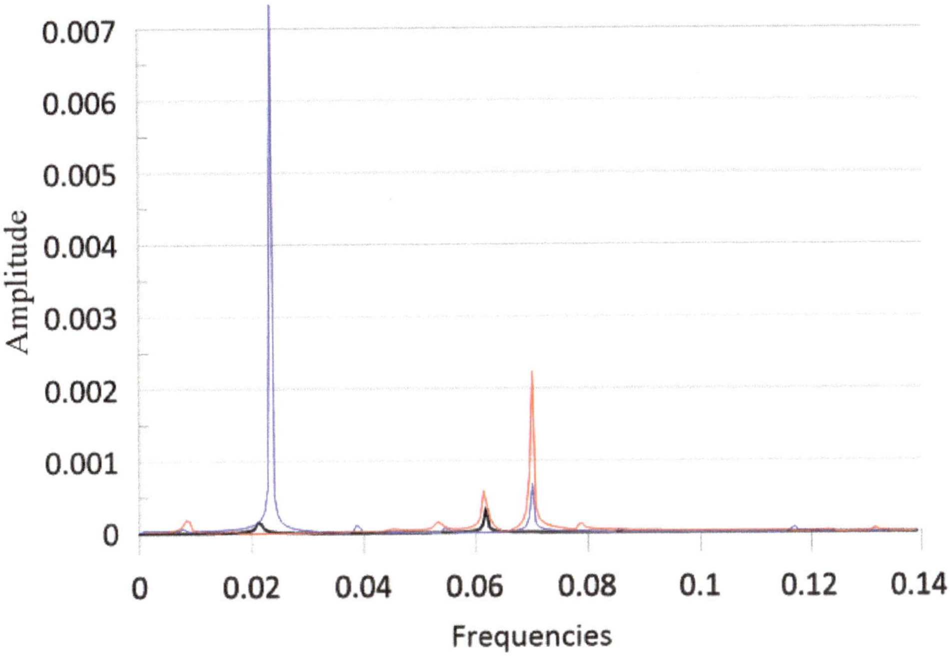

Figure 17. FFT of the responses of the time series plotted in Figure 15. $T_{up} = 15$, $T_d = 10$; blue line: $T_{up} = 26$, $T_d = 10$; red line: $T_{up} = 35$, $T_d = 10$. $f_1 = 0.1 \times 10^{-5}$ and $\omega = 0.07$.

7. Applicability of One-Mode Reduction

As mentioned above, the use of the one-mode reduction model would provide significant advantages in studying the nonlinear dynamics of a beam, if this model can be validated. The examples below show the applicability of this model as well as its limitations.

In Figure 18, the results obtained by the 1-DoF model, using the same data as those used to obtain Figure 4a by the 3-DoFs model, are shown. As one can see, the results are in perfect agreement. Obviously, for these low levels of load amplitudes and $\Delta T = 0$, the 1-DoF model is enough to obtain correct results.

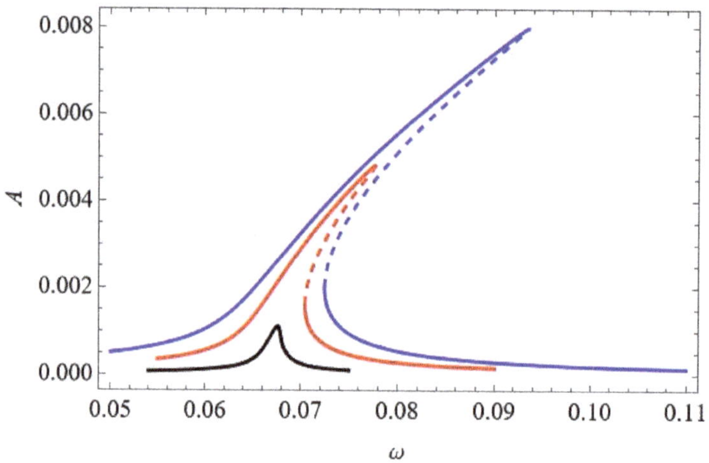

Figure 18. Resonance curves around the first natural frequency for small mechanical loading: f1 = 0.1×10^{-6}—black; f1 = 0.5×10^{-6}—red; f1 = 0.1×10^{-5}—blue; $\Delta T = 0$.

The results shown in Figure 19 correspond to the case presented in Figure 5. The amplitude of the loading is much bigger than in the previous case. This leads to essential changes in the resonance curves obtained by the different models. The amplitudes of vibrations in the case of the 1-DoF model are larger, but more significant is the fact that the 3-DoFs model predicts instability of the upper branch of the resonance curve. This cannot be obtained by a 1-DoF model.

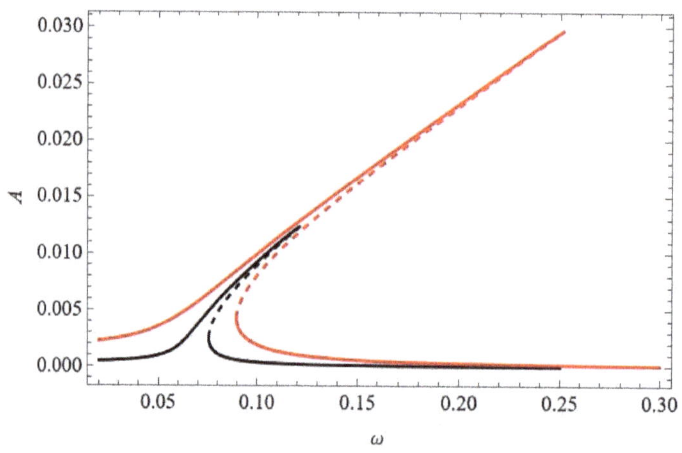

Figure 19. Resonance curves around the first natural frequency for large mechanical loading: $f_1 = 0.2 \times 10^{-5}$—black; $f_1 = 0.1 \times 10^{-4}$—red; $f_2 = 0$; $f_3 = 0$; $\Delta T = 0$.

Most of the curves shown in Figure 20, plotted for different temperatures, are very similar to those shown in Figure 7a. However, for $\Delta T = 15$, the additional instability obtained by the 3-DoFs model cannot be predicted. Of course, the interaction of modes and partially stable and partially unstable isolated solutions are obtained only by the 3-DoFs model for higher temperatures, and the 1-DoF model cannot obtain them.

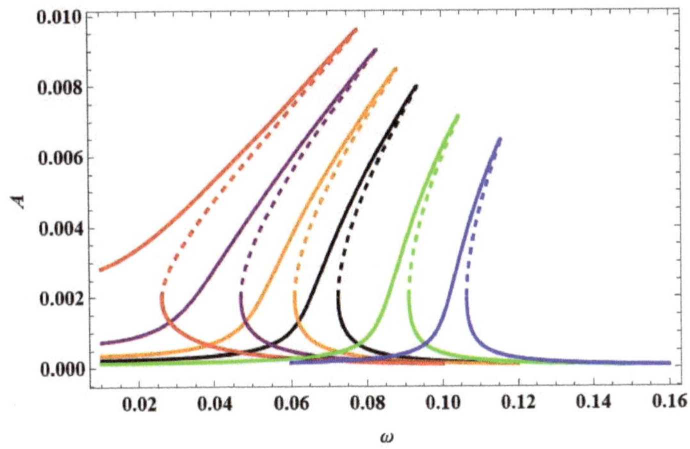

Figure 20. Resonance curves around the first natural frequency for mechanical loading $f_1 = 0.1 \times 10^{-5}$ and different values of elevated temperature: $\Delta T = 0$—black; $\Delta T = 5$—orange; $\Delta T = 10$—magenta; $\Delta T = 15$—red; $\Delta T = -10$—green; $\Delta T = -20$—blue.

If one compares Figure 21 with Figure 8, the same conclusion can be made—multiple solutions ($\Delta T = 15$) and many instabilities in the solutions cannot be detected by the 1-DoF model.

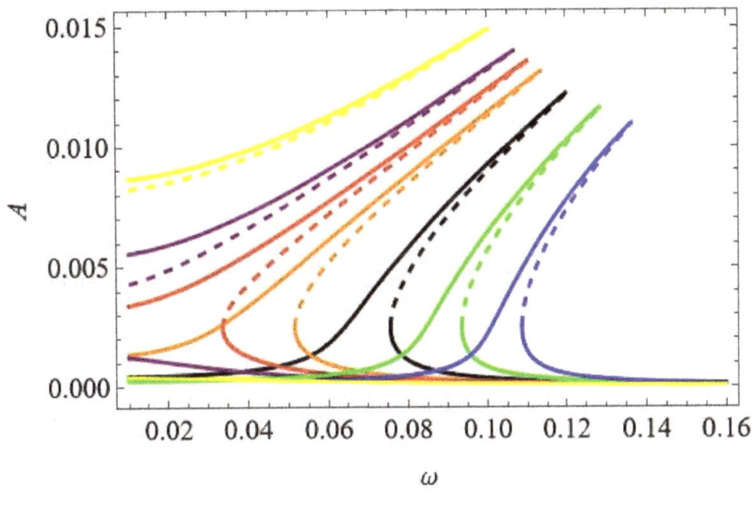

Figure 21. Resonance curves around the first natural frequency for mechanical loading $f_1 = 0.2 \times 10^{-5}$ and different values of elevated temperature: $\Delta T = 0$—black; $\Delta T = 10$—orange; $\Delta T = 15$—red; $\Delta T = 20$—magenta; $\Delta T = 30$—yellow; $\Delta T = -10$—green; $\Delta T = -20$—blue.

The bifurcation diagrams in Figure 22 show similar behaviour of the beam response q_1 as the one in Figure 9a, but the values of the maximal amplitudes are different, and some instability in the upper branches of the curve for the maximal temperature is not detected by the 1-DoF model.

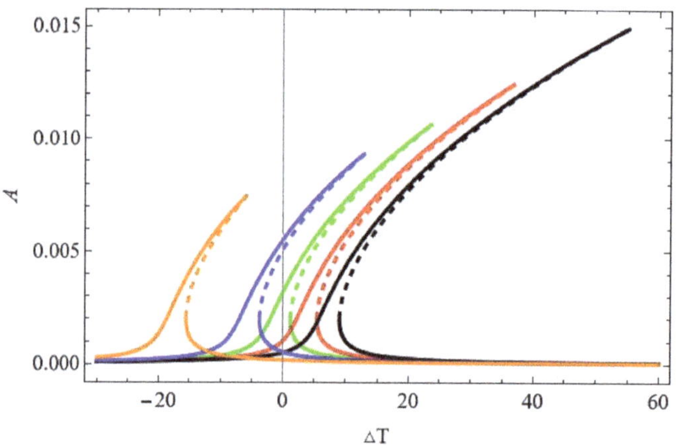

Figure 22. Bifurcation diagram of the beam response against elevated temperature ΔT in the vicinity of the first natural frequency and fixed mechanical loading $f_1 = 0.1 \times 10^{-5}$ and different excitation frequencies: $\omega = 0.05$—black; $\omega = 0.06$—red; $\omega = 0.07$—green; $\omega = 0.08$—blue; $\omega = 0.10$—orange.

In the case of a linear distribution of the temperature along the beam thickness, the results obtained analytically are similar to those shown in Figure 12a. The asymmetry due to the geometric and material properties of the layers is also confirmed analytically.

8. Conclusions

In this study, a nonlinear thermoelastic model of a bimaterial beam is developed, considering mechanical and thermal boundary conditions. The model enables the study of the influence of mechanical and thermal loading, considering the heat linearly distributed through the beam thickness. The model is validated through comparison with finite element analysis, including vibration modes, frequencies, and time response under periodic excitation for clamped–clamped boundary conditions.

The impact of mechanical and thermal loading is investigated for the three-degrees-of-freedom model and then compared with the analytical solutions obtained by the harmonic balance method from the reduced one-degree-of-freedom model. The models with one and three degrees of freedom exhibited strong agreement for relatively small or moderate periodic forces, demonstrating that higher vibration modes (second and third) have a minimal effect on the system's behaviour. The resonance curves showed stiffening, particularly at larger amplitudes. However, for large-amplitude oscillations of the bimaterial beam, an instability zone occurs in the frequency response curve because of a strong interaction of the second and third vibration modes, resulting in additional solutions. The reduced model based on one mode cannot detect this phenomenon. Moreover, the increased temperature leads to bifurcation and bimaterial beam buckling with asymmetric post-buckling oscillations and strong modal interactions. Then, even four stable solutions may exist with small- or large-amplitude oscillations around each of the two buckled states, and also, the isolated solution of a large amplitude with oscillations between two buckled states may occur. The shift of the resonance curves and a nonlinear stiffening effect is observed in the case of linearly distributed temperature throughout the beam thickness. However, if the sign of the temperature gradient is changed, the shift of the curve is different. By increasing the temperature of the upper beam surface and fixing the temperature of the bottom, a stronger hardening phenomenon is obtained compared with the case of elevated temperature. The bifurcation point occurs at a higher temperature, and thus, post-buckling dynamics are shifted as well. However, the beam response is in qualitative agreement with the response for the beam under elevated temperature.

In conclusion, the analysis underscores the importance of considering mechanical and thermal effects, particularly at elevated temperatures, to accurately predict the dynamic behaviour of bimaterial beams under periodic loading. The findings highlight the significance of higher vibration modes, temperature distribution, and post-buckling dynamics, contributing valuable insights into the design and performance of such systems.

The obtained results demonstrate the possibility of transitioning to more complex chaotic oscillations, and this problem will be investigated by a detailed analysis of bifurcation diagrams, basins of attractions, Poincaré maps, Lyapunov exponents, and Fourier transformation in future studies, considering the coupling of mechanical and thermal fields.

Author Contributions: Conceptualization, E.M. and J.W.; methodology, J.W.; software, E.M.; S.D. and J.W., validation, S.D., J.W. and E.M.; investigation, S.D., J.W. and E.M.; writing—review and editing, S.D., J.W., and E.M.; supervision, J.W. All authors have read and agreed to the published version of the manuscript.

Funding: This work was accomplished with financial support from the Bulgarian research fund, grant KP-06-N72/7, 2023, partially supported by the Centre of Excellence in Informatics and ICT under the Grant No BG16RFPR002-1.014-0018, financed by the Research, Innovation and Digitalization for Smart Transformation Programme 2021–2027 and co-financed by the European Union. The first and third authors acknowledge this support. The research of the second author was funded in part by the National Science Centre, Poland, Grant No. 2021/41/B/ST8/03190.

Informed Consent Statement: Not applicable.

Data Availability Statement: The original contributions presented in this study are included in the article. Further inquiries can be directed to the corresponding author.

Conflicts of Interest: The authors declare no conflicts of interest.

Nomenclature

x	Longitudinal coordinate
z	Transverse coordinate
u	Displacement in longitudinal direction
w	Displacement in transverse direction
t	Time
l	Length of the beam
b	Width of the beam
z_m	The coordinate of interface layer
$h^{(i)}$	Thickness of ith layer
$h = h^{(1)} + h^{(2)}$	Thickness of the beam
$\psi(x,t)$	Angle of rotation of the normal to the mid-surface of the beam
ε	Strain vector
$\varepsilon_x^0 \varepsilon_{xz}^0$	Components of strain in the mid-axes
κ_x^0	Curvature in the mid-axes
k^2	Shear correction factor, depending on the geometry; for a rectangular section, taken as $k^2 = 5/6$
$S = \{\sigma_x, \sigma_{xz}\}$	Stress vector
$\sigma_x^{(i)}$	Normal stress in ith layer
$\sigma_{xz}^{(i)}$	Tangential stress in ith layer
$E^{(i)}$	Young's modulus of ith layer
$G^{(i)}$	Shear modulus of ith layer

$\alpha_T^{(i)}$	Coefficient of linear temperature expansion of ith layer
$\overline{Eh} = E^{(1)}h^{(1)} + E^{(2)}h^{(2)}$	
$\overline{Gh} = G^{(1)}h^{(1)} + G^{(2)}h^{(2)}$	Generalized flexural beam's rigidity
$\overline{EI} = \frac{1}{3}\left\{E^{(1)}\left(z_1^3 + \frac{h^3}{8}\right) + E^{(2)}\left(\frac{h^3}{8} - z_1^3\right)\right\}$	
T	Difference between the temperature of the beam and the ambient temperature
T_d, T_{up}	Temperature on bottom (down) and upper beam surfaces
$p(x,t)$	Transverse loading
c_1, c_2	Damping coefficients
d_1, d_2	Dimensionless damping coefficients
$W_i(x)$	Normal-mode shape of beam transverse displacement
$\Psi_i(x)$	Normal-mode shape of beam rotation angle
$q_i(t)$	Generalized coordinate, time function
ω_n	Natural frequencies of the linear elastic undamped beam
P, ω	Amplitude and frequency of harmonic excitation
ξ_n	Modal damping coefficient

Appendix A

The coefficients in Equation (34) are obtained by using Equations (27), (31), (33), and (35). For the particular case considered in the paper, they have the following numerical values:

$C_{(1)111} = 0.81681 \times 10^2$	$C_{(1)222} = -0.62987 \times 10^{-2}$	$C_{(1)333} = -0.51910 \times 10^3$
$C_{(1)112} = -0.50505 \times 10^{-2}$	$C_{(1)113} = -0.19388 \times 10^3$	$C_{(1)122} = 0.30558 \times 10^3$
$C_{(1)223} = -0.24185 \times 10^3$	$C_{(1)133} = 0.75817 \times 10^3$	$C_{(1)233} = -0.13175 \times 10^{-1}$
$C_{(1)123} = 0.48940 \times 10^{-2}$		
$C_{(2)111} = -0.16836 \times 10^{-2}$	$C_{(2)222} = 0.11432 \times 10^4$	$C_{(2)333} = -0.17481 \times 10^{-2}$
$C_{(2)112} = 0.30556 \times 10^3$	$C_{(2)113} = 0.24461 \times 10^{-2}$	$C_{(2)122} = -0.18894 \times 10^{-1}$
$C_{(2)223} = -0.24399 \times 10^{-2}$	$C_{(2)133} = -0.13175 \times 10^{-1}$	$C_{(2)233} = 0.24536 \times 10^4$
$C_{(2)123} = -0.48346 \times 10^3$		
$C_{(3)111} = -0.64642 \times 10^2$	$C_{(3)222} = -0.81442 \times 10^{-3}$	$C_{(3)333} = 0.52654 \times 10^4$
$C_{(3)112} = 0.24469 \times 10^{-2}$	$C_{(3)113} = 0.75800 \times 10^3$	$C_{(3)122} = -0.24184 \times 10^3$
$C_{(3)223} = 0.24532 \times 10^4$	$C_{(3)133} = -0.15566 \times 10^4$	$C_{(3)233} = -0.52363 \times 10^{-2}$
$C_{(3)123} = -0.26341 \times 10^{-1}$		
$D_{111} = -0.28030 \times 10^{-3}$	$D_{112} = 0.57776 \times 10^{-8}$	$D_{113} = 0.22185 \times 10^{-3}$
$D_{121} = 0.57775 \times 10^{-8}$	$D_{122} = -0.10486 \times 10^{-2}$	$D_{123} = 0.74706 \times 10^{-9}$
$D_{131} = 0.22183 \times 10^{-3}$	$D_{312} = 0.74705 \times 10^{-9}$	$D_{133} = -0.22502 \times 10^{-2}$
$D_{211} = 0.51769 \times 10^{-7}$	$D_{212} = -0.10670 \times 10^{-11}$	$D_{213} = 0.40972 \times 10^{-7}$
$D_{221} = 0.10670 \times 10^{-11}$	$D_{222} = -0.19366 \times 10^{-6}$	$D_{223} = 0.13797 \times 10^{-12}$
$D_{231} = 0.40969 \times 10^{-7}$	$D_{232} = 0.13797 \times 10^{-12}$	$D_{233} = -0.41559 \times 10^{-6}$
$\omega_1 = 0.63799 \times 10^{-1}$	$\omega_2 = 0.17570$	$\omega_3 = 0.34402$

References

1. Boley, B.; Weiner, J. *Theory of Thermal Stresses*; Wiley: New York, NY, USA, 1960.
2. Nowacki, W. *Thermoelasticity*, 2nd ed.; Pergamon Press: Oxford, UK, 1986.
3. Thornton, E.A. *Thermal Structures for Aerospace Applications*; American Institute of Aeronautics and Astronautics: Reston, VA, USA, 1996.
4. Udrescu, R. Aerothermoelastic behaviour of skin panels exposed to alternating thermal excitation. In Proceedings of the European Congress on Computational Methods in Applied Sciences and Engineering ECCOMAS 2000, Barcelona, Spain, 11–14 September 2000; pp. 1–11.
5. Mei, C.; Abdel-Motagaly, K.; Chen, R. Review of nonlinear panel flutter at supersonic and hypersonic speeds. *Appl. Mech. Rev.* **1999**, *52*, 321–332. [CrossRef]
6. Travis, L.; Turnery, Z.; Mei, C. Finite element analysis of the random response suppression of composite panels at elevated temperatures using shape memory alloy fibers. In Proceedings of the 35th Structures Structural Dynamics, and Materials Conference, Hilton Head, SC, USA, 18–20 April 1994.
7. Shi, Y.; Lee, R.Y.Y.; Mei, C. Thermal postbuckling of composite plates using the finite element modal coordinate method. *J. Therm. Stress.* **1999**, *22*, 595–614. [CrossRef]
8. Hao, H.; Scola, C.; Semperlotti, F. Flexural-mode solid-state thermoacoustics. *Mech. Syst. Signal Process.* **2021**, *148*, 107143. [CrossRef]
9. Tang, Y.; Ding, Q. Nonlinear vibration analysis of a bi-directional functionally graded beam under hygro-thermal loads. *Compos. Struct.* **2019**, *225*, 111076. [CrossRef]
10. Tang, Y.; Zhong, S.; Yang, T.; Ding, Q. Interaction Between Thermal Field and Two-Dimensional Functionally Graded Materials: A Structural Mechanical Example. *Int. J. Appl. Mech.* **2019**, *11*, 1950099. [CrossRef]
11. Yeh, Y.-L.; Chen, C.-K.; Lai, H.-Y. Chaotic and bifurcation dynamics for a simply supported rectangular plate of thermo-mechanical couplingin large deflection. *Chaos Solitons Fractals* **2002**, *13*, 1493–1506. [CrossRef]
12. Tiwari, R.; Abouelregal, A.E.; Shivay, O.N.; Megahid, S.F. Thermoelastic vibrations in electro-mechanical resonators based on rotating microbeams exposed to laser heat under generalized thermoelasticity with three relaxation times. *Mech. Time-Depend. Mater.* **2024**, *28*, 423–447. [CrossRef]
13. Abouelregal, A.E. Computational analysis of thermoelastic vibrations of functionally graded nonlocal nanobeam excited by thermal shock. *J. Vib. Control* **2023**, *30*, 3105–3116. [CrossRef]
14. Saetta, E.; Rega, G. Unified 2D continuous and reduced order modeling of thermomechanically coupled laminated plate for nonlinear vibrations. *Meccanica* **2014**, *49*, 1723–1749. [CrossRef]
15. Saetta, E.; Rega, G. Third order thermomechanically coupled laminated plate: 2D nonlinear modeling, minimal reduction, and transient/post-buckled dynamics under different thermal excitations. *Compos. Struct.* **2017**, *174*, 420–441. [CrossRef]
16. Settimi, V.; Saetta, E.; Rega, G. Nonlinear dynamics of a third-order reduced model of thermomechanically coupled plate under different thermal excitations. *Meccanica* **2020**, *55*, 2451–2473. [CrossRef]
17. Settimi, V.; Rega, G.; Saetta, E. Avoiding/inducing dynamic buckling in a thermomechanically coupled plate: A local and global analysis of slow/fast response. *Proc. R. Soc. A Math. Phys. Eng. Sci.* **2018**, *474*, 20180206. [CrossRef] [PubMed]
18. Settimi, V.; Saetta, E.; Rega, G. Local and global nonlinear dynamics of thermomechanically coupled composite plates in passive thermal regime. *Nonlinear Dyn.* **2018**, *93*, 167–187. [CrossRef]
19. Settimi, V.; Rega, G. Asymptotic formulation of the nonlinear bifurcation scenarios in thermomechanically coupled plates. *Nonlinear Dyn.* **2023**, *111*, 5941–5962. [CrossRef]
20. Yeh, Y.L. Chaotic and bifurcation dynamic behavior of a simply supported rectangular orthotropic plate with thermo-mechanical coupling. *Chaos Solitons Fractals* **2005**, *24*, 1243–1255. [CrossRef]
21. Huang, K.; Zhou, X. Analysis of linear free and forced vibrations of microbeams with thermoelastic damping. *Eur. J. Mech. A/Solids.* **2025**, *112*, 105672. [CrossRef]
22. Warminska, A.; Manoach, E.; Warminski, J. Dynamics of a circular Mindlin plate under mechanical loading and elevated temperature. *MATEC Web 400 Conf.* **2016**, *83*, 05013. [CrossRef]
23. Warminska, A.; Manoach, E.; Warminski, J. A reduced model of a thermoelastic nonlinear circular plate. *MATEC Web Conf.* **2018**, *148*, 06001. [CrossRef]
24. Manoach, E.; Warminska, A.; Warminski, J.; Doneva, S. A reduced multimodal thermoelastic model of a circular Mindlin plate. *Int. J. Mech. Sci.* **2019**, *153–154*, 479–489. [CrossRef]
25. Shang, Y.F.; Ye, X.Y.; Feng, J.Y. Theoretical analysis and simulation of thermoelastic deformation of bimorph microbeams. *Sci. China Technol. Sci.* **2013**, *56*, 1715–1722. [CrossRef]
26. Carpinteri, A.; Pungo, N. Thermal loading in multi-layered and/or functionally graded materials: Residual stress field, delamination, fatigue and related size effects. *Int. J. Solids Struct* **2006**, *43*, 828–841. [CrossRef]
27. Carpinteri, A.; Paggi, M. Thermo-elastic mismatch in nonhomogeneous beams. *J. Eng. Math.* **2008**, *61*, 371–384. [CrossRef]

28. Srinivasan, P.; Mark, S. Spearing Effect of Heat Transfer on Materials Selection for Bi-material Electrothermal Actuators. *J. Microelectromechanical Syst.* **2008**, *17*, 653–667. [CrossRef]
29. Pazera, E.; Jędrysiak, J. Effect of Temperature on Vibrations of Laminated Layer. *Vib. Phys. Syst.* **2018**, *29*, 2018032.
30. Manoach, E.; Warminski, J.; Kloda, L.; Warminska, A.; Doneva, S. Nonlinear vibrations of a bi-material beam under thermal and mechanical loadings. *Mech. Syst. Signal Process.* **2022**, *177*, 109127. [CrossRef]
31. Timoshenko, S. *Strength of Materials (Part I)*, 2nd ed.; D. Van Nostrand Company: New York, NY, USA, 1948.
32. Hjelmstad, K.D. *Fundamentals of Structural Dynamics*; Theory and Computations; Springers Nature: Cham, Switzerland, 2023.
33. Krack, M.; Gross, J. *Harmonic Balance for Nonlinear Vibration Problems (Mathematical Engineering)*, 1st ed.; Springer: Berlin/Heidelberg, Germany, 2019.
34. AUTO: Software for Continuation and Bifurcation Problems in Ordinary Differential Equations. Available online: http://indy.cs.concordia.ca/auto/ (accessed on 20 June 2025).
35. ANSYS. Southpointe 2600 Ansys Drive Canonsburg, PA 15317 USA. Available online: www.ANSYS.com (accessed on 20 June 2025).

MDPI AG
Grosspeteranlage 5
4052 Basel
Switzerland
Tel.: +41 61 683 77 34

Materials Editorial Office
E-mail: materials@mdpi.com
www.mdpi.com/journal/materials

www.ingramcontent.com/pod-product-compliance
Lightning Source LLC
Chambersburg PA
CBHW062039220926
51850CB00010B/2600
* 9 7 8 3 7 2 5 8 6 6 8 6 1 *